D0374591

JUDGING A BOOK BY ITS LOVER

ALSO BY LAUREN LETO

Texts from Last Night: All the Texts No One Remembers Sending
(with Ben Bator)

JUDGING A BOOK
BY ITS LOVER

*A Field Guide to the Hearts and
Minds of Readers Everywhere*

LAUREN LETO

HARPER PERENNIAL

NEW YORK • LONDON • TORONTO • SYDNEY • NEW DELHI • AUCKLAND

HARPER ⬤ PERENNIAL

JUDGING A BOOK BY ITS LOVER. Copyright © 2012 by Lauren Leto. All rights re-served. Printed in the United States of America. No part of this book may be used or reproduced in any manner whatsoever without written permission except in the case of brief quotations embodied in critical articles and reviews. For information address HarperCollins Publishers, 10 East 53rd Street, New York, NY 10022.

HarperCollins books may be purchased for educational, business, or sales promo-tional use. For information please write: Special Markets Department, Harper-Collins Publishers, 10 East 53rd Street, New York, NY 10022.

FIRST EDITION

Designed by Lisa Stokes

Library of Congress Cataloging-in-Publication Data is available upon request.

ISBN 978-0-06-207014-2

12 13 14 15 16   OV/RRD   10 9 8 7 6 5 4 3 2 1

CONTENTS

III. HOW TO FAKE IT

IV. SNARK BAIT

AUTHOR'S NOTE

LET ME BE PLAIN when I state that my judgments, wisecracks, and sarcastic comments come from a place of deep admiration for every one of the authors whose work I discuss in these pages. There is nothing more beautiful than a well-written book, and there is nothing more admirable than the attempt to create something beautiful.

The Justice League of Ex-Teachers of Mine

THE FIRST BOOK I ever loved was a book about a monster in a child's closet. I had a hard time learning how to read when I was in first grade. I remember feeling overwhelmed by the fields of letters, the spaces, the punctuation. I have a clear memory, however, of being brought into the hallway one day by my teacher. She opened a book and walked me through, slowly, how to string everything together and follow, sentence by sentence, a cohesive story. And it was a garish story. The kind of story that children's book authors seem pathologically drawn to: a kid is utterly terrified by a monster, who, inexplicably—when the kid finally tries to talk to the monster—turns out to be friendly. *Easy enough*, I thought, and continued on rereading that same book every day for the rest of first grade.

In second grade, lightning struck when my teacher told me I was good at reading. If you tell an eight-year-old she has a talent for something, she'll never give it a rest; you tell her, "Oh, you're funny!" and the child will keep making raspberries and pretending to be a monkey until you want to rip her arms out. You say, "Wow! You're pretty good at basketball," and it will end in tears as you finally pull him by his hair off the court. Tell her, "Hey! You're a great singer!" and you're in for it—you'll get a flouncing bundle emitting unbridled music-like monstrosities whom you'll have to stab in the heart before it'll be quiet ever again.

My astute parents had up to this point avoided telling me I was good at anything. Our nightly dinners were spent reinforcing the message "It's not funny" to my brother, for good measure repeatedly reminding my sister she should "stop drawing everywhere," and me, my directive was to "stop imagining things." But Mr. Booker, that unsympathetic man, complimented me one day on my reading skills. It was all downhill from there. When my parents figured out what the teacher had done, they marched into the principal's office to complain. "She does nothing but read!" "She has no friends!" "Her nose is always in a *book*!" Meeting only silence and bewilderment there, they picketed the PTO: "It can't be healthy! Reading *all the time*!" "What if she needs glasses someday?! *Boys don't make passes at girls who wear glasses!*"

Unfortunately for them, I had been told I was good at reading and I was not going to stop. I read in my closet with all the lights off and nothing but a flashlight, finally coming out only after my mom had, in an abrupt state of panic, called the police. I read in the bathtub. In the process I used up all the hot water because a leaky drain necessitated that the water be left running for the tub to stay full. My brother and sister took years of ice-cold showers because I couldn't put down anything from *Animorphs* to Nabokov (I still do this—they've asked me to stop coming home for the holidays). "Go outside and make some friends," my mom would say with a sigh. I'd try to please her by Rollerblading in a circle in our driveway while reading. Unsurprisingly I didn't make many friends that way. I refused to ride my bike to school because there was no way to read while biking. Instead I'd walk in an ambling fashion with my face in a book, resulting not once but twice in my somehow arriving home with only one shoe on. "This needs to stop! You have a problem!" my father cried. He sent me to summer camps to straighten me out but I'd hide contraband in my suitcase and spend the week away from home in Narnia.

Teachers in subsequent grades would complain to Mr. Booker, that innocent man, "If you had just kept your mouth shut she'd be listening in class instead of hiding her face behind a book!" They quickly banded together, the Justice League of Ex-Teachers of Mine, in order to throw a side-eye at anyone who dared encourage children, lest they turn out like me. They met every Monday morning, to start the week off right, in the teacher's lounge. The password to get in was "mediocre"—a key element of their mission statement: "Keep them mediocre, keep our jobs easy." They'd sit and stew in there. "I've run out of stickers to put on her reading chart. This child is a drain on the system; my sticker budget has run dry!" They'd join my parents at the PTO picket. STICKER HOG! their signs cried out.

By the time I was in high school, the attendance list had grown so large they had to move the meetings to the gym. "Hi, I'm Ms. Washington and I was Lauren's teacher in tenth grade. One time, I had to send her home because she was crying about losing her annotated copy of *Catch-22*. Children shouldn't have to miss class because they're sad about a book. It's disgusting and subversive." "Hi, I'm Mr. Young and I was Lauren's teacher in ninth grade. Once, she turned in what was supposed to be a book report but instead amounted to a terrifyingly detailed account of J. D. Salinger's life. I am sincerely worried she might be stalking him." "I'm Mr. Montague and I was Lauren's twelfth-grade teacher. I was unable to get her to stop laughing when *Lear*'s Earl of Gloucester dives off the cliff. It confused the other students and I had to give her detention."

Upon graduation, no one was sad to see me go: "She's somebody else's problem now!" "I bet she doesn't finish college, she'll be too busy reading." "You can't while away all your time reading in college!"

PART I

For the Love of Print

Commercial Confessions

I'M AN ANXIOUS PERSON. My guess (gathered from an unscientific survey of fellow readers and the uneducated opinions of my family) is that this may be the result of years of overexposure to fictional worlds and underexposure to real-world activities such as recess, school dances, and cocktail parties. I'm not very comfortable in settings and situations most people take for granted as part of the comings and goings of everyday life. For example, traveling: traveling with me is an experience I wouldn't want to wish on anyone—and I go to great lengths to save friends and family the trouble. Accompanying me on planes and in cars is nightmarish. If it weren't for the helpful tricks that I've come to rely on, I don't know how I'd get anywhere. I've developed ways to deal with my anxiety, tics that keep the pressure down and keep the terror at bay. These quirks are my dirty little secrets. Sometimes it's just two stiff drinks at the airport TGIF before boarding; other times the situation calls for more drastic actions to divert my attention from my mounting anxiety over the prospect of hurtling forward on a road or through the sky. I need something a little more potent.

I'm telling you this because I want to be as honest as possible with you. Janet Evanovich books are my booze; I can't board a plane without checking the airport bookstore to find the newest tale of Stephanie Plum. If I've read all the available Evanovich, I have to pick the next-easiest, sleaziest thing. I started and finished *Twilight* on red-eye trips from Detroit to Los Angeles and back; I conquered *New Moon* before touchdown from New York to San Francisco. I wept over Idaho while reading the first *Hunger Games*. At these moments I need my reading easy and quick; I need to turn the pages without knowing it. I don't have the bandwidth to wonder about the underlying meaning of the exact word chosen to phrase how one turned around or analyze just why an object was described in a certain way. I need distraction, not deep thoughts.

I make this distinction because most of this book is about avoiding bad books, and I don't want a reader to think I'm being an elitist snob. Considering yourself a serious reader doesn't mean you can't read light books. Loving to read means you sometimes like to turn your head off. Reading is not about being able to recite passages from Camus by memory. Loving young adult novels well past adolescence isn't a sign of stunted maturity or intelligence. The most important thing about reading is not the level of sophistication of the books on your shelf. There is no prerequisite reading regimen for being a bookworm.

Let's all embrace the fact that *The Da Vinci Code* has sold more copies than all of Saul Bellow's works combined. Dan Brown and his ilk are keeping our bookstores in semi-business. If America chooses easy escapism over dense dialogue, we should welcome that decision with equanimity. When it comes to reading, whatever floats the boat. And if someone deems your reading choice frivolous, who cares? If it's what you want to read, go for it.

However, silly books shouldn't be all we read. We have to acknowledge that there is a problem with an exclusive diet of the latest hot commercial fiction and nonfiction: after a while, you realize, the books blend together. The voices, the stories, the characters, the arcs of the drama—after a while, it can all start to feel so . . . familiar. If we get too comfortable in our reading choices—too lazy—we're giving something up. Kids get turned off of reading before they even begin in earnest because they recognize the predictability of it all. Die-hard readers who stick to Nicholas Sparks must have missed a few steps on the road through adolescence. How does one waste significant time reading and never open a book by Philip Roth? Before a middle schooler reads another "boy meets girl" young adult novel, we should hand them a copy of *Snow Crash* by Neal Stephenson or sit them down with a Brontë sister. Small steps to open up their perspective on plots a bit.

Nabokov described great stories as "supreme fairy tales"—they take our imagination to work, igniting dreams we wouldn't know how to express in our daily lives. The best books expand and challenge the mind. The "easy" books don't give us folds and symbols to look beneath or around; they don't have images that come to you suddenly when you're alone on a street corner and a passing man's face suddenly strikes you, like the line in Jeffrey Eugenides's *Middlesex*, as "rumpled like an unmade bed." They don't have perfectly captured vignettes that live on for you beyond the book and enter your life; like the line from Lorrie Moore's *Self-Help* that comes to you as you accept a third date with someone you really could see yourself with, just like the last guy: "You love once, I told you. Even when you love over and over again it is the same once, the same one."

I am not a scholar of literature. All my commentary comes from my experiences and presumptions as a reader. As you read what follows—my one-sentence book reviews, my gross generalizations about others' bookshelves, my categorical statements on how to fake any author, my love lavished on little-known treasures in literature, my cheat sheets on how to write like any author, my horror stories from my life as a bookworm, and my open letters to authors' fans— feel free to be annoyed if I snark on an author you love. Feel free to berate my schooling—my college degree is in political theory and constitutional democracy, a pretentious way of explaining that at one point in time I read a lot of Dostoyevsky and Plato. I'm a law school dropout and I managed to fail my college precalculus math class three semesters in a row. I'm afraid to get on planes. I am not an authority. I'm a Janet Evanovich fan, for Christ's sake.

The Bookshelf of the Vanities

WHO AM I TO comment on bookshelf displays? My family home has a bookshelf filled with . . . pottery. But it wasn't always like that. I grew up in a home with the most beautiful bookshelf you've ever seen. My father spent weeks constructing a built-in bookshelf that spanned the entirety of a long wall in our living room, save a center spot for the television. The bookshelf was painted a bright white and had intricate molding, contrasting with the raspberry-colored walls and deep-green carpeting of our living room. For the better part of a decade, I turned to those shelves to find my mom's extensive collection of Kitty Kelley biographies, *The Andy Warhol Diaries*, Ken Follett books, past yearbooks, and Stephen King novels.

A few years ago my mother decided the bookshelf was no longer for holding a book collection. Instead, it would be for her burgeoning pottery collection. I visited my family's home over the holidays to find, instead of rows of books, a neatly arrayed series of bright greenish-blue ceramic pottery lining the living room wall. Plates, candlesticks, vases, all in various shades of aquamarine, which my mother swore up and down matched with her green carpet and raspberry wallpaper in some way my untrained eyes couldn't perceive.

Where were the books? She had relocated them to a low cabinet at the end of the room, near our kitchen. The cabinet had a broken latch, so the door swung open violently no matter how steadily and slowly you tried to crack it. Instead of being neatly lined up, the books were placed in plastic bins according to no method that I could discern. Harry Potter found himself in the same crate as Norman Mailer, my father's guide to self-employment next to my brother's *Calvin and Hobbes*. Tears sprung to my eyes as I saw Edgar Allan Poe's *Tales of Mystery and Imagination* piled underneath numbers five and nine of Janet Evanovich's Stephanie Plum series. The bins could hardly fit in the cabinet, so every time I opened the doors, they would lurch forward at me. I'd use one hand to push the bins of books back, a knee to hold the cabinet door steady, and my free hand to root through and find what I wanted.

I felt I had to set the books free. I had to rescue them from my mother's shuttered hiding place. So I mailed the ones I wanted to protect to myself in New York City. With no doorman at my fifth-floor walk-up apartment to receive packages, I would have to go to the post office, pick up one box at a time (there were four in total), and walk the fifteen blocks home with—by that time—arms that felt like they

were slowly catching fire. Then I'd head up the five flights
and unpack the books onto the small coffee table that I fash-
ioned into a type of bookshelf by lining up rows between
the legs and across the top. My next apartment was also on
the fifth floor. I had a friend help me with the boxes this
time, our arms burning for days afterward. There, I lined the
assorted paperbacks and hardcovers across a pair of drawers,
stacking one row on top of the other to fit them all. In a
third apartment, I bought two cheap ladders from a hard-
ware store and stacked books across their rungs, the ladders
leaning against the wall. It was improvisation. I didn't have
the money to invest in a choice piece of furniture but dis-
playing my books felt like an important step of moving in.
Ask anyone with a big book collection, and they'll tell you
moving them was the hardest part of the move. Take down
a bookshelf and there's often no less than four, possibly up
to eight, good Lord if it's over ten, boxes of dense material.
This is the single greatest argument for welcoming e-books.
Abandoning print and having your Kindle on display instead
doesn't sound like such a bad idea while carrying book box
number seven to the car.

For some reason, even people who don't read own
bookshelves or a sort of showcase for their books. They
seem to be a necessary component of "the home," an adorn-
ment no self-respecting adult can live without. But what,
when you really look at them closely, do these simple pieces
of furniture tell you when they're filled with their owner's
library—books received as gifts, bought in an airport on the

outgoing leg of a vacation, or idly picked up while in line at a store? You are in someone's home. Somewhere—likely right in front of you—is a wealth of information about who your host is, or who they want you to think they are. Let's review a few of the typical bookshelf presentations you may encounter and the personalities behind them.

The Tortured Artist

Story: You have a passion for denying yourself a steady job and a resolve to keep posting YouTube videos until you get a viral hit (but so far views haven't tipped past the double digits).

Books: Charles Bukowski (heavily read and marked up), Milan Kundera (read heavier in the areas where the protagonist describes his apathy toward commitment), Nietzsche (attempted, abandoned).

Objects: Bottle of Maker's Mark strategically propping up the Bukowski, some change, rolling papers.

Bookshelf: Windowsill just within reach of your mattress on the ground.

Sorority Girl

Story: You are marked by an inability to recognize yourself as a caricature of what the media thinks twenty-year-old girls act and think like.

Books: Chelsea Handler (any book by her, bought at an airport bookstore), *Eat, Pray, Love* (but you've never heard of *Committed*), Jack Kerouac (a memento from high school days).

Objects: Pendant with Greek letters; oversized, painted martini glasses; pink frame with picture of four or more overly made-up women; Andy Warhol poster of Marilyn Monroe.

Bookshelf: The cubby in the upper part of your desk, built into the bunk at your sorority house.

Fraternity Guy

Story: Books aren't your main concern—in fact they don't even rank in the top one hundred things you care about—but it's college and your mother bought you a shelf on a parent's-weekend shopping spree, so why not?

Books: Tucker Max (received at your high school graduation party; you have an involuntary memory as to where specific stories are located in the book), *The Bachelor's Guide to Survival* (again, a gift from your high school graduation), *Between a Rock and a Hard Place* (unread).

Objects: Empty beer can, girl's necklace, speakers.

Bookshelf: Ikea-brand that your parents helped you put together; will be discarded or simply left in place instead of taken home after graduation.

Quirky Hipster

Story: You've run the gamut from Gabriel García Márquez to Miranda July. It is your sense of adventure that drives you to find the next most obscure author to champion. Either that or your pretentiousness.

Books: Che Guevara biography (purchased from independent bookstore), Susan Orlean's *The Orchid Thief*, copies of *The Paris Review* (lined up just right), graphic novel (unread gift from ex-boyfriend).

Objects: Ceramic owl statue, cardboard cutout of a unicorn, vintage typewriter.

Bookshelf: Shelves ordered from Etsy and placed with loving precision on your bedroom wall.

Brash Entrepreneur

Story: You read all the business books and take notes so you can be sure to apply their priceless lessons with your team. You don't appreciate the irony of taking up your time to read books revolving around the philosophy of "get things done by sitting around and reading about how to plan to get it done."

Books: Tim Ferriss's *4-Hour Workweek* and *4-Hour Body*; *The Starfish and the Spider*; *The Checklist Manifesto*.

Objects: Kitschy piggy bank, KEEP CALM AND CARRY ON poster, clock, buckyballs.

Bookshelf: Books are placed on your minimalist-style glass desk.

Old-Money Prep

Story: Contemporary literature is of no interest to a man or woman who defines themselves by their history.

Books: *Stuff White People Like* (ironically), *Moby-Dick*, Ayn Rand, *Mrs. Dalloway*.

Objects: Wooden frame with dated family photo, wooden sailboat.

Bookshelf: Heavy wood, antique—like at an upstate cabin.

Brooding Academic

Story: Initially unable to let go of graduate school, you are transitioning from your tortured-artist phase by gradually acquiring a taste in novels with stronger roots in philosophy, since they're so much fun to debate with your friends.

Books: Fyodor Dostoyevsky, William Gaddis, *The Master and Margarita*.

Objects: Legal pad with outline for a screenplay hidden under the first two pages.

Bookshelf: Doesn't matter, as long as it can hold all the books you own and show them off with intimidating effect.

Soccer Mom

Story: You are too busy to mentally invest yourself in a substantial novel but enjoy the often sedative effects of the perfunctory *O, The Oprah Magazine* book of the month.

Books: Biography (whatever presidential autobiography has been released recently), Jodi Picoult, Jennifer Weiner.

Objects: Pottery, photos of children, other assorted trinkets and artifacts from the suburbs.

Bookshelf: A section of the entertainment system in your living room.

Dorky Dad

Story: Once a fraternity brother or drummer in a garage band, now you're thinking about mortgage payments and kids' soccer games instead of hitting on chicks or expressing your inner self.

Books: John Grisham, the *Da Vinci Code* series, David Baldacci (but you only read fiction on your annual Easter trip to Florida), tennis and/or golf instruction books, a couple old college textbooks.

Objects: Bulky address book, decorative mug from Father's Day, leather sofa nearby.

Bookshelf: A large bookshelf made with the best wood—the highest-quality option in the catalog. Located in your study.

That Certain Bookstore Smell

WALKING INTO A GREAT bookstore elicits a powerful emotion. You're flanked floor-to-ceiling by the spines of books, surrounded by tables decked out with new releases; everywhere your eye lands you're suddenly aware of the pages and pages of histories, stories—sheer information—and the impossibility of ever getting to read it all. Front tables display every element of great, carefully curated bookstores you can ask for: the handwritten notes about books chosen by such and such employee and a one-or-two-line description; artful covers that could double as wall art; annoying movie-version covers of greats. You can while away your time indefinitely in a bookstore. For most of us, the ability to wander the aisles without a specific purpose is priceless. And there is nothing better than a bookstore with a courteous staff who knows when to help you and when to leave you alone.

After dropping out of law school I spent a year trying

(unsuccessfully) to write my first novel and (a bit more suc-
cessfully) running a humor blog. To say a lot of that year
was also spent without a solid daily agenda would be an
understatement. Every day I'd arrive at a small office that
I rented and I'd clack away at my laptop keys until I hit a
block. Luckily, across the street lived a Borders bookstore,
so I'd jet over there to procrastinate and try to invigorate my
writing by osmosis from the immortals. I was addicted to
that store. To those of you who scoff at the likes of Barnes &
Noble and (the lately demised) Borders, what are called the
big-box chains—spread out all across our country—I make
no excuses for my Borders addiction; I lived in a small town,
and that was the only store available to me. But, given the
choice, I'll take the independent over the chain on principle.

You've Got Mail, the Nora Ephron–scripted flick star-
ring Meg Ryan and Tom Hanks, was always confusing to
me. In it, Tom Hanks plays a big-shot executive of a branch
of discount bookstores and Meg Ryan plays the owner of a
small, independent children's bookstore. Their paths inevitably intersect as their businesses go head-to-head. Eventually, Ryan's small store goes under and the discount store
flourishes. For some reason, Ryan falls in love with Hanks
afterward. I'm not going to start in on the holes in a Nora
Ephron plot, but I've always been puzzled by the fact that
this romantic comedy lets the big guy win. Hanks's slick
stores are portrayed with a kind of superiority, as if the film
is embracing them as the obvious choice over Meg Ryan's
devoted little enterprise. Hanks's character compares the

books in his store to *cans of olive oil*. Every time I see the movie, I want to go visit my closest independent bookseller and buy out the front table. You'll rarely meet a better-read or more passionate person than your local indie bookseller. There is nothing greater than a highly curated bookstore. There's something essential about the scene—a mess of a table full of phenomenal titles and no distractions in the checkout line other than more rows of books. You don't get that at a Barnes & Noble. They're hardwired to hawk movie tie-in editions and the latest bestseller by Dr. Oz.

But I'll admit I'm no purist, and I'm not innocent. Tom Hanks's character predicts he'll win over the neighborhood because his store's "books are cheap and [the store has] cappuccino." Sometimes we all fall victim to convenience. If I need to pick up a tchotchke for someone's birthday, maybe I'll grab a coffee as well. And while I'm at it, why not grab the latest Neal Stephenson? Before you know it you're breezing by the Michael Crichton stepladder on your way up the fifth-floor escalators of your Barnes & Noble megaplex.

Recently, the Borders in my old neighborhood closed. I'd since moved away to New York City but the feeling that I'd lost the place where I went to lose my time in my hometown during adolescence hit me harder than I expected. It was the only bookstore, independent or otherwise, in my city. I'd met a boyfriend there, discovered Dostoyevsky there, spent money I didn't have there on the two-for-the-price-of-one tables. I realized early on that the Bantam mass-market paperbacks were reliably the cheapest editions of the classics, with their tissue paper and small print. I set up my study groups in their café, my business partner and I discussed how to create a website in their humor section, I bought my Mother's and Father's Day gifts there, I bought my nonreader girlfriends *Why Men Love Bitches* for all their birthdays. I bought my reader friends Kerouac, Vonnegut, Maugham. I attended the book signing of a young girl who had self-published a novel at the age of twelve. I was camped out in the children's book section when I received an e-mail from a friend telling me Salinger had died.

People often talk of the smell of bookstores, a concentrated fragrance of paper and bindings. But there's more to it than a certain smell; there's a humming excitement in the air that, if you think about it, seems out of place for a room filled with objects that cannot twirl, bleat, or shine. Bookstores contain the residue of thousands of people who went in there to find an experience, a narrative that guided them to a new place or reinforced what they were doing. Whether you find yourself in Meg Ryan's or Tom Hanks's store, there's always a quiet corner and a new story to find.

Ten Rules for Bookstore Hookups

"ARE YOU LAUREN?" SAYS the guy kneeling in front of the Eastern philosophy section, having just plucked a book off the shelf as he gets to his feet. It's the *I Ching*, now resting, spine cracked, in his hands. I consider saying no. He's cute but I've been down the Eastern-philosophy-boy path before and I have no intention of returning. To make matters worse, I'm holding a Thomas Friedman book. "Lauren . . . Leto?"

Oh shit, he knows me, I think. "Hi! Yes." I own up to my identity. It turns out he's a coworker of one of my friends.

We get into a discussion about why we're there and what we're buying. I explain that I'm not a Thomas Friedman fanatic but was interested enough in *The World Is Flat* to take a peek. He explains that he doesn't have an interest in Eastern philosophy but there's a conductor he's been listening to quite a bit who's a real *I Ching* devotee, so he thought he'd check it out. I exhale a bit.

We dated for years after that first meeting. So, I feel I can

propose these ten rules for bookstore pickups with a measure of authority.

First and most importantly: it's all about the books. The
best thing about meeting someone in a bookstore is that
they are surrounded by things to talk about and are possibly
holding the very nugget that you can use to weasel your way
into their heart.

1. Admitting ignorance of any given author or book is
   no huge strike against you, but at least have a solid
   exit strategy for any talk that's over your head. If
   you're stuck, feigning an obsession with another, more
   obscure author is usually the best way to go. "No, I've
   never really been able to get into Bellow—when all my
   friends were reading him I was pretty preoccupied with
   Bernhard."

2. Explain away an otherwise embarrassing book in hand
   by saying it's a gift or required reading for a book club.

3. Having a great conversation? Go with it. Try to stroll
   around the aisles, pointing out your favorites or new
   releases you're anxious to get to. Going really well?
   Transition into the bookstore's coffee shop or one
   nearby (there's always one nearby). Most of the time
   that's just a matter of a few steps through the magazine
   aisle. Can't get any easier than that.

4. For a laugh check out the self-help section for love.
   If for some reason they don't think the deluge of
   information on how to snag a mate is funny, run.

5. There's an endless amount of things to talk about while
   browsing the store: their favorite books in high school,
   why beautiful book covers entice us so much, how
   The Virgin Suicides is the most often shoplifted book
   in America, anecdotes like how Alain de Botton spent
   a week in Heathrow as the self-proclaimed "writer
   in residence" and got a book deal out of it (he turned
   the experience into *A Week at the Airport*, in case you
   were wondering). See how many interesting turns the
   conversation can take from that?

6. It's easy to begin a conversation. Sidle up to the guy
   or girl dismantling the Russian lit section. "What're
   you looking for?" (make sure it's obvious you're not an
   employee). Smile. He says, *"The Master and Margarita."*
   You respond, "My friend was just telling me about that
   book! They said it was great." Don't get in over your
   head. Inside a bookstore is not the place to go toe-to-toe
   with a reader by bluffing.

7. People reveal a lot about themselves just through
   their browsing behavior. We're creatures of habit and
   chances are high if someone is sitting in the history
   section, they probably would love whatever historical

biography you most recently read—you don't have to reach far. Start the conversation by suggesting a good book along the lines of the one they're already holding. Best-case scenario, they've already read it. There could be hours of conversation material there, if you want it. Point is, it's low-hanging fruit. It's like people are wearing signs announcing what their passions are.

8. Sometimes a long checkout line is best for the challenge. Chances are less likely you'll be able to solidify a real connection with someone in the ten minutes or less it takes to move through the line, but hey, chemistry takes only a split second to establish, so try it out.

9. Try to get through the whole experience without asking the cute guy or girl where they work or live. The best part about picking someone up in a bookstore is that you don't need to know anything about their personal life in order to have an amazing conversation.

10. Bring up the rise of e-readers. Do they own one? Do they like it? Talk about your favorite bookstore in whatever city and how you couldn't imagine the store not existing anymore. Lament the fact that in a world without bookstores, you would never have met.

Rules for Public Reading

THEY'RE BEAUTIFUL THINGS, tangible books. The iPad, Nook, and Kindle are swiftly taking away our ability to instantly judge people by their choice of reading material in public places, but for a little while longer, you'll be able to strike up a conversation with a stranger, or silently mock them, as you notice them cracking open a wonderfully bulky copy of *I Am Charlotte Simmons*. In the age of print, you didn't think twice about lugging a book around on your quotidian errands— not as a measured public display, but simply as a fact of daily life. A burden assumed willingly so you have something to do while you wait at the doctor's, something to preoccupy you while waiting for a friend at a restaurant, or in case you have a few free minutes to get back to the chapters you tore yourself away from to leave for work.

There are times when you can't deny that your choice of one book over another on your way out the door has something to do with your destination. A voice in your ear tells you to forgo *Angela's Ashes*—you don't want to have tears streaming down your face in the waiting room unless you want to attract concerned looks from nearby patients. Off for a drink? Brandishing *House of Holes* at a bar may be sending the wrong (or right) message, if you're entertaining any hopes of being picked up.

Here are some tips for your reading venue of choice, when what title you're holding in your hands is public information.

Universal Guidelines, for Waiting Rooms, Subways, and Everything in Between

Don't be awful. There's nothing cute about reading *Twilight* in public. Save that for nights alone, when you realize you've been single for far too long and there's no end in sight. Don't exaggerate emotions and never laugh out loud while reading. Furrowed brows or the hint of a smile are sometimes acceptable, but a beaming mug is creepy. Why? You are not in the privacy of your living room, and you should be capable of harnessing some powers of self-control.

Train or Subway

Big-city living offers paradoxically comforting and anxiety-inducing public transportation. Always best to get through the commute with a book in hand. Sure, you can go stereotypical with a novel by a liberal author who received a favorable review in the *Times*—and who is therefore also likely a white man. To fit in, *Let the Great World Spin* or anything by Gary Shteyngart (but God help you if you're trying to get through *The Russian Debutante's Handbook*). But stereotypes need not be adhered to so stringently. Read what you love; you don't want to get stuck underground with a bad book.

Park

People will think you're a psycho if you read William Gaddis in the park. Your reading material can be serious, but it should also be brief. Camus's *The Stranger* or Eugenides's *The Virgin Suicides*, for example. Or a paperback small enough to hold with one hand but whose subject matter and title are congruent with your demeanor as you sit squinting in the sun, peering into the book as if deciphering a message from outer space.

Plane

People love to claim that anything goes on a plane. You're stuck in a hermetically sealed container with crying babies and coughing grandmothers—why not treat yourself to an effortless Grisham novel? Have you seen airport bookstores? Tucker Max's latest is forever on the front-and-center tables. I want you to think about this for a second. If the plane were to crash, do you really want a Jennifer Weiner book found in the clutches of your charred hands? (Yes, they will probably find an Evanovich in my hands, but I've made my peace with that.) If you're going to die in a flash of insane pain, at least allow yourself the pleasure of peering down at a dreamy author like Michael Chabon or Nicole Krauss. Watch out for the prospect of parasite readers, glancing over your shoulder as you're going through the (very many and graphic) sex scenes in *True Things About Me* by Deborah Kay Davies. Nothing more uncomfortable than the look in an elderly stranger's eyes after they notice you're reading about a woman being tied up and done anally.

Beach

Read Peter Benchley's *Jaws*. Read *Gentlemen Prefer Blondes* by Anita Loos. Ladies: go ahead, be light and be breezy, but don't carry chick lit unless you're already in a committed relationship. No guy or girl is going to pick up someone reading *Something Borrowed*. Men: stay away from *Freedom* or the latest presidential memoir. Nothing says "I only read on vacation" like the latest It book on the *New York Times* bestseller list.

Coffee Shop

Anything goes. You're surrounded by scones, wooden tables, and the sweet nectar that has helped make every writer into an author (alongside whiskey—if it's even necessary to mention). It's even fine to publicly annotate in a coffee shop. They're created for reading and writing. One caveat: I'm speaking about *independent* coffee shops. Starbucks and their ilk are for meetings with people you don't care to meet with and finishing term papers. No comfort can be found in their too-small tables and wobbly chairs.

Bar

I once sat at a bar next to a man reading Charles Bukowski. He struck up a conversation with me about how my night was going; I thought he was very nice. Except he kept waving his book around and fidgeting with it. After a while, my friend leaned over and said, "I think he wants you to ask him about his book." So I did.

"How's that Bukowski?"

"Oh, this?" he said. "I love to drink whiskey at this bar while I read Bukowski. It's inspiring." He paused to open up the book to where he left off, revealing marginal notes and underlined phrases, then closed it again, with evident pride. "When I'm at a bar, living a life like he lived, I feel that I'm making him proud."

I nodded.

"We're very similar," he said, continuing.

"That's nice," I said as I contemplated exit strategies for this conversation. See, Bukowski is great. And drinking while reading Bukowski is actually a requirement, so I understand that point. But would I pick up a guy at a bar who touts his similarity to Bukowski? Hell no.

You can read any of the hard drinkers at a bar: Joyce, Fitzgerald, etc. And picking someone up at the bar while reading one of those authors? Stellar idea. In so doing you are attracting a crowd with an equal appreciation for the solace to be found in a scotch and a sordid story line. I'd avoid reading any of the latest and greatest books because you'll get a bunch of "Oh, I heard that was good" or an "I just finished that for my book club." Also, never read business books at a bar. No one wants to sleep with Mr. or Mrs. Productivity.

Survival of the Nerdiest

"I LIKE TO READ a lot on the weekends," the female says, her hair piled high in a way that signals sophistication.

The male, leaning forward—a sure sign of interest—responds, "Oh, I haven't read a novel in a while. I've been meaning to read [insert name of book most recently at the top of the *New York Times* bestseller list, a J. D. Salinger novel, or a book most recently made into a blockbuster movie], though."

Here we have a basic snapshot of a reader's mating ritual. If we listen closely to these two readers in the wild, we can learn how subtle cues and choices reveal the eligibility of the potential partner. Unknowingly and almost instantly, the male of the species has shown his hand and exposed his unsuitability as a mate.

Suitors who cite high-profile books as their most likely next read in a bid to impress are clearly stunted in their development. By choosing a book at the top of the *New York Times* bestseller list, they aim to convey their intelligence and their healthy acquaintance with literature, when mistakenly, they have done just the opposite. This is the classic move of any nonreader attempting to mack on a reader, male or female. Field tests show they are usually college educated, outgoing, and middle-class.

Then there are the specimens who seem to think that J. D. Salinger is a byword for literary savvy and aloofness. They'll trot out *A Catcher in the Rye*, thinking that mere mention of the title will signal their quietly pained but soulful alienation from the pack. They're trying to signal they've been out of the book game for a while and they're about to reenter with gusto, particularly if it means gaining the reader's attention. It rarely ends up actually happening. The Salinger come-on appears to be in use by American males aged twenty to thirty years (I'm speaking from hard-earned experience). When confronted with evidence of a lady's bookishness, the nonreader fumbles and mentally grasps for the one book they believe to be an accepted part of the canon. The reader, on the other hand, has settled the score with Salinger at the appropriate time—adolescence—and long ago moved on. The reader took Salinger in stride as a balm for her adolescent angst but has emerged from that period of her life a full-grown adult.

When lobbing back and forth latest releases in the midst of the courtship ritual, the worst possible move is to prattle off the title of a book recently adapted into a movie. This mistake also seems to be among the most widespread. Not only do potential mates unwittingly employ this misguided technique, but practically every nonreader who has reason to make small talk with a known reader likes to play this card. These are likely the same individuals who, if for some godforsaken reason they find themselves moved to follow through on their threat, are sure to pick up the movie tie-in edition, revamped with airbrushed portraits of the stars in character. Unfortunately for them, Jude Law doesn't actually make an appearance in the novel and grabbing a book because you like the actors in the movie version rarely correlates with success.

Oh, the Men You'll Love

IF I WERE WRITING a book about the culture of music fandom, this is where I'd recount my experience dating a reckless, ne'er-do-well bassist in a punk band. If I were writing about movies, I'd share stories about my stormy affair with an over-enthusiastic indie filmmaker or a self-obsessed actor. Since I'm writing about books, you're probably expecting anecdotes about my pursuit of aspiring authors, the MFA grads headquartered at the corner table of my local coffee shop, the adjunct college professors lurking around trivia night at my local bar. Maybe even an adoration of the young, attractive author who just had a smash debut release. But—and maybe this will surprise you—I don't have any such stories to tell. I'll admit to my share of crushes. What I wouldn't do to meet the boyishly handsome Simon Rich is a short list. It's hard not to wonder, while paging through a novel, about the discussions you could have with the comely figure on the back cover photo. Or how wonderful the role of muse could be with the brooding man at the bookstore around the corner. But, when it comes down to it, my real fascination isn't with anything so simple as just a fiction writer. I've found my love to be for a state of mind, not a profession.

I can't resist contrarians, mercurial enigmas in pursuit of an intellectual dispute. Devotion to me means Friday nights debating politics, Stieg Larsson's rape scene, the music of Skrillex, or anything else that prompts our ire at the bar, and weekend mornings throwing the paper in anger over Thomas Friedman's latest overly metaphorical article. The fighters, the truth seekers, the champions of deductive reasoning and considered argument, the enemies of idiocy, or the friends of derision, depending on how you look at it.

They say you're supposed to love men like your father. The only evidence I have of captiousness in my father's kind heart is his advice to my younger self after I'd asked him whether he rooted for the University of Michigan or Michigan State University (a distortion of syntax I found difficult to believe could distinguish two separate, rival institutions). He chose my future alma mater, Michigan State University, because he claimed to "always root for the underdog"— which I suppose is a sort of pitying contrariness. Other than that, my dad is an even-keeled, "don't rock the boat," solid sort of man.

I've been hooked on hotheads ever since the boy in my second-grade class who insisted on going by his full name—first, middle, mother's maiden, father's surname— argued with my third-grade teacher over the validity of girls being permitted to wear their Blossom-style hats indoors while boys were forced to remove their caps. This point of contention had arisen over *my* Mayim Bialik–esque hat, as a matter of fact. I was the girl who was making a mockery of fairness. The boy held forth before the whole class every time our teacher told him to remove his hat, demanding to know why the rules didn't apply to me. He would train his finger on me for the entirety of his presentation. Spencer Tracy's monologues in *Inherit the Wind* would give me flashbacks to that pointed, accusing finger forever after. Our teacher responded, impertinently, that the hat looked "cute" on me, and suddenly I found myself taking up arms with the boy. I stood and countered with, "So, you're saying an ugly girl couldn't wear this hat?" Hot tears filled my eyes. Eight years old and I was already a bleeding heart, feeling the pain of the more ungainly girls. The boy was chagrined by the collusion of his unlikely ally. He'd intended to keep the argument about gender, not about perception. Or maybe, like so many other men I've loved since, he just pivoted his stance to stay safely outside (and above) the bounds of others' arguments. "No, no. She's saying that boys aren't cute. All girls are cute," he said, correcting me (I said the men I loved were argumentative, not necessarily logically sound). The dispute was finally settled when our teacher suggested that those who chose to question her reasoning might want to stay after school to discuss it in greater detail, but before that I had already removed my hat, in ceremonious and imagined solidarity with my unsightly sisters.

I find a lot of my fellow book fanatics are attracted to the same type. The saying may be "Opposites attract," but in our world it seems oppositions are attractive. A disputing nature is a deal maker instead of a deal breaker. It's an affinity that has its origins, perhaps, in our early appreciation for what words can do. While kids on the playground were getting riled up about sports and music, we had our heads in books. We carried around the objects of our desire wherever we went, willing to leave ourselves open to ridicule. We developed sagged, rounded shoulders from heavy backpacks. And, of course, the glasses we'd inevitably need became cherished attributes of our allegiance to words. But we also developed a capacity for self-defense. Argument. Self-defense with words. And you learn how to recognize partners in this struggle. While we're immersed in Evelyn Waugh's fictional worlds, infatuated with Mr. Darcy, enthralled by Sinclair Lewis's *Babbitt*, without knowing it, we are also acquiring an appreciation for the way dialogue flows between characters.

Words, the way we learn to bend and twist letters into guardrails for whatever logic we're trying to prove. Read heavily since childhood and you become scarily good at doing this; we tend to learn it through osmosis, never setting out to become masters of debate but suddenly finding ourselves in ninth grade and standing up at our desks in heated words with another classmate over why Mark Twain's use of the N-word in Huckleberry Finn shouldn't have been edited out of future editions. And then we sit down, exhilarated by skills we hadn't realized we possessed. Who better to set your eye on than the very man you were just arguing with? All those romantic comedies, with the clipping banter and sarcastic snarks between the pair before they finally realize they're in love with each other, make sense to us. Woody Allen is one of us.

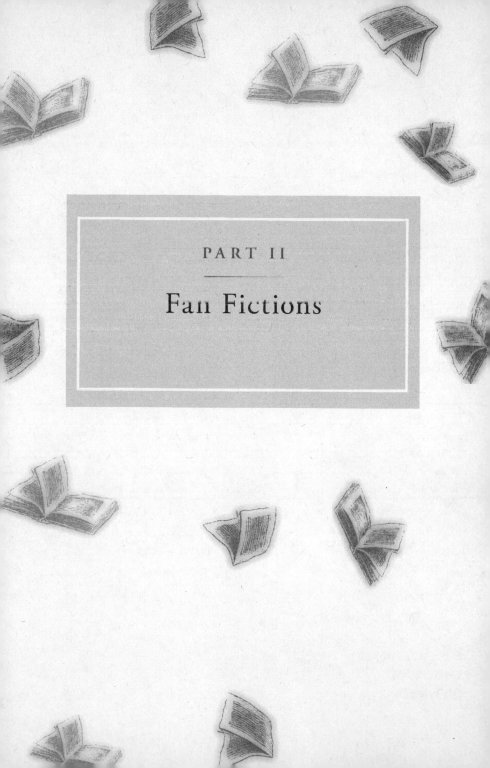

PART II

Fan Fictions

Harry Hardships

WHEN I WAS YOUNGER, I used to make wish lists for upcoming holidays, twenty books deep. Christmas and my birthday were the only times my parents would contribute to my book collection. Otherwise, I'd get movies or a tennis racket. Maybe a jacket, best-case scenario. "Oh no no. There won't be any Kerouac. We'll get you a movie and then *maybe you'll find some friends* to watch it with."

For my thirteenth birthday my mother bought me a Harry Potter book.

I cried.

I argued she didn't respect my intellect. That she didn't appreciate the ease with which I had picked up Shakespeare in honors English that year. I very likely told her that she was stifling me as a person and trying to dumb me down by purchasing books for dumb children instead of *The Bell Jar*, like I had asked for.

"You don't understand me!"

"You're right. I don't," she said, shrugging. My mother was used to "not understanding" me. Her theory was that her inability to understand me was due to her being born without an imagination, while I had been born with such a large one. As if she couldn't figure out what pants to put on me because she was petite, whereas my butt was big. She couldn't carry on conversations with my junior-high self because she never inhabited, or desired to inhabit, fantasy worlds.

I put the Harry Potter book in the "stupid books" section of my bookshelf and resumed my consumption of what I saw as more fitting material.

Somehow, one bored night, with my Christmas stock running low, I found myself with nothing to read. So I tipped the book off the shelf, dusted it, and scanned the first few pages in the interest of self-distraction. Over a decade later, I'm a twenty-four-year-old who dresses up as Harry Potter characters for midnight showings, who owns a poster mug shot of herself with UNDESIRABLE #1 written under it (a present from a friend), who will argue against any Harry detractors, no matter how intimidating. Against my better senses, I fell hard and fast into the fan vortex of J. K. Rowling's unassuming hero.

During sophomore-year finals in college, an e-mail was sent to all the kids in my program about a paid opportunity to teach English and creative arts in Japan. Before I even responded to the e-mail, I called my older brother to tell him I was going to be spending my summer in Japan. Born a year apart, my brother and I were similar in almost every way and we hated each other for it. There's never been one opportunity he's had that I haven't coveted and vice versa. We took the same high school math class; we were in the same ski classes, swimming lessons, and safety courses. Due to an ill-conceived casting method involving names pulled out of a hat, we were once slated to appear as Mary and Joseph in our church's nativity play, and my brother haughtily bowed out after my refusal to defer the role for a year, so Peter might shine on his own, unmarried to his sister. I even graduated college one year early so I could be right there, enrolled and sitting next to him, on his first day of law school. We had a little sister, born five years after me and adored by my parents more than the two of us put together, our show pony years having passed long ago by the time she entered the world. We were competitive not to be last, and not to be first, so as not to be considered overeager.

My brother's addiction to video games rivaled my own to books. No other destination on the globe would have excited my brother, as a gamer, like Japan. I knew this when I called him, shoving my trip in his face like we were ten years old again.

"You haven't been accepted yet," he told me.

"I know, but I'm going to get it."

"When is it supposed to be?"

"The month of July."

"Ha!" he yelled, and I realized my error too late. "You're going to miss when *Harry Potter* comes out!"

The last Harry Potter book was due to be released at the

end of July. Another long-standing rivalry between the two of us revolved around who loved Harry more; the one who knew how the series ended first would surely be the winner of this competition, regardless of any handicapping factors, like geographical location.

The swirl of theories concerning the end to the Harry Potter series was at full pitch that May. I would sit on the computer in Harry Potter–themed forums, under user names along the lines of Laurenione13 or Slythereto, laying out theories for communal dissection. I'd critique the hypotheses of others: "No way is Draco going to make any sort of altruistic gestures during the final battle," and "The prophecy applies to Neville, it makes so much sense."

I ended up getting the job in Japan. My preemptive boasts provided the sorority girls I was then living with convenient evidence in support of the very trendy Oprah philosophy of The Secret. I'd be responsible for guiding a class of fourteen-year-old students in the writing and performance of a play, all in English. I arrived thrilled at the thought of taking my pupils through a modern redo of an Ibsen play or a melodramatic enactment of something vaguely Shakespearian. These were smart, well-to-do Japanese kids. I was sure they'd be more brilliant than I.

For the majority of the first class, I played word games, attempting to get a sense for their grasp of the language before we started writing the play. We went in a circle and I had everyone name (in English) a national capital until we were all out of countries. They then suggested American eighties bands as a category and, after that, American movies. They were more familiar than me with these artifacts of American culture. I was relieved at their genius, not realizing the pop culture theme of their suggestions was a red flag.

Afterward, a gaggle of girls from the class approached me. They held stapled-together sheets of paper in their hands and looked nervous. One spoke up: "Leto-san, we were wondering if you would let us act our script." In unison they held up their papers.

"You already have a script for the play?" I asked, as if it possibly could be for something else, like the thirteen-year-olds wrote commercials and screenplays on the side. They nodded excitedly and one girl handed me her copy.

She explained as I read. "It's a fashion show . . . and there's a beautiful girl. Then Jack Sparrow comes and tries to steal her! So her boyfriend fights him! And the boyfriend wins!"

"Jack Sparrow?"

"*Pirates of the Caribbean!*" all seven girls screamed excitedly at once. That's how I ended up producing a fan-fiction play in Japan.

On the day the new Harry Potter book came out, two of the American guys I worked with wanted to get the book as well. The more geographically minded teacher led the way. He had a map with our route from Ohanajaya to Roppongi (the area containing the nearest store selling the English-language version of the book, we were told) planned out. We had to transfer three times, taking four trains in total to our destination. The beginning of the trip was without complications, save for the dirty looks we got from other teachers when we announced we were skipping out on work early that day to buy a children's book. Some might argue that the Harry Potter series is not for children, but it is. Death and destruction aided by simple language and a lack of sex is child's play.

After the second transfer, we made it through the turnstile just in time to hear the train arrive. We ran for it, the two boys' long legs outpacing mine. I was two steps behind when they reached the closing doors. Japanese trains aren't lazy like the subways of New York; when they close, they close. No exceptions. The boys made it through the doors and I stood watching them through the glass window.

I panicked, screeching, "What do I do?!" They tried to yell back but we couldn't hear each other through the glass. They resorted to anxiously gesturing at me, making an "O" on one hand and holding four fingers up on the other. I couldn't make out what they were doing. "Oh-four?! Oh-four?!" The train sped away and I burst into tears. I was alone in a place where I couldn't speak the language. I didn't even know the name of the station where I was standing, I had forgotten the name of the station we were going to, and I didn't know the name of the station we had departed from. I wondered if I should just stand there, forever. Become a bum, lost in the outskirts of Tokyo, foraging for food or English speakers.

Most of all, I worried the boys were going to get the last two available Harry Potter books and by the time I made it, if I ever made it at all, the store would be sold out. Then I definitely wouldn't finish it before my brother. After some moments, which felt like hours, a man tapped my shoulder and started speaking Japanese to me. I shook my head over and over as he spoke louder and louder, repeating a phrase. In hindsight, he was doing the same thing Americans do while speaking to foreigners who don't know the language: acting as if language barriers can be overcome if you speak loud enough. Finally, he started miming out the same actions as my friends, an "O" with one hand, four fingers up on the other. I said back, "Oh-four? Oh-four?" The man kept shaking his head and had just about lost his patience when a woman standing nearby ran to the map and started gesturing for me to come look. Forty was the number of the stop she was pointing to. "I go there?" I said as loud as the man had been talking to me, pointing to myself, then at the map. She and the man nodded. All three of us got on the next train when it arrived and the two practically shoved me out the door when it arrived at stop 40, where my friends were waiting for me, doubled over from laughing at my difficulties.

We finally arrived in Roppongi and I got the book, reading the entire thing that night, depressed because Neville didn't get to be the hero but elated that I finished it before my brother.

The Rules of Book Club

THE BOOK CLUB IS one of the most joyful and most annoying things in a reader's life. With every meeting, there is the one awkward moment when everyone becomes aware of who didn't read the book or who didn't understand the protagonist's state. Don't be that person.

Book clubs tend to have many required roles; there's usually a self-appointed leader who holds the group together, gives the deciding vote for the next book the group will read, and decides where to meet. Next, there's the guy who talks too much about the wrong ideas. He's probably a philosophy major from some elite college— too bad the tuition didn't buy the ability to realize when you're monopolizing a conversation and need to shut it. Don't forget the girl who sits and stares like a mute. Maybe she's bored on Monday nights and just shows up for the hell of it. Maybe she's too shy to give her opinion. Either way she's not going to say a word. Last and worst, however, is the nodder. The nodder is the one who starts, yes, to nod frantically while you're halfway through explaining a point. When especially excited, the nodder also likes to murmur positive reinforcement like, "Yes, exactly," as if you needed a chorus to affirm your explanation of why you think Owen Meany knew from the beginning that his life had a greater purpose.

Here's an explanation of what usually goes down at book club, presented in the style of the rules of fight club from the book *Fight Club* by Chuck Palahniuk.

Rules of Book Club

1. The first rule of book club is you have to talk about book club. Outside of book club, any mention of a book you read in book club must be accompanied by, "Oh, I read that in book club." The café where you meet for book club must be similarly noted in conversation: "Oh, I go there for book club." The people in your book club are not friends; you'll refer to them in phrases like, "Oh, we're in book club together." Further examples: "I can't go to brunch on Sunday, I have book club." "Yes! I'm so excited, we chose the book I wanted for book club." "I don't know if I have time to join your book club, I'm already in three." "I'm going to go see that movie with my book club because we read the book in book club."

2. The second rule of book club is you *have to* talk about
   book club. "Ever since we read *C* for book club, I can't
   stop reading Tom McCarthy! I'm suggesting we read
   his book *Men in Space* for next book club. I think our
   book club needs to focus on him again; we really
   accessed some good points about him and ourselves,
   as a group." "I brought up our relationship problems
   in book club and they said if I stayed with you it'd be
   like Patty staying with Walter in *Freedom*, which is a
   book I read for book club, so I think we should break
   up." "While in book club today I thought about how
   my life seems so much like Sloane Crosley's in *I Was
   Told There'd Be Cake*, which is the book we read for book
   club, so I'm quitting my job."

3. If someone says stop, goes limp, taps out, the fight is
   over. "I can't continue fighting with you over whether
   or not Jennifer Egan had a solid finish to *A Visit from
   the Goon Squad*. It's clear you took her 'autistic chapter'
   as definite proof of her transcendent creativity when I
   thought it wasn't an adequate setup for the ending of
   the book, and I think a robustly resolved plot better
   serves the reader than do quirky structural devices.
   It went on for too long and the closing chapter did
   nothing to resolve what went on with Bennie and
   Sasha. We can sit here all day and say that was the
   purpose of her book, that there are no clear finishes to
   anyone's story in life, but I refuse to accept that, given

how she so clearly demarcates Lulu's life. Everyone else was left open-ended except for Lulu, and I don't understand that but I have to give up since you think strewn-together pie charts make for resonant storytelling."

4. Two guys to a fight. Have you ever heard three or more people fight over a story line in a book club? It's a mess. The fight needs to be kept to two, or else other people come in out of left field with completely irrelevant statements in order to hear their own voices. Keep it to the two debating and the others can sit and make notations until they clear it up.

5. One Brontë sister for the life of the book club. And no Austen. You should've quenched your thirst for the sentimental novel in high school.

6. You must have annotations and reading glasses. How are we going to know you read without notes in the margins? How are we even going to know you can read without reading glasses? Reading glasses are the mark of years of abusing your eyes, and we want proof you've been through the wringer before you sit at our table.

7. Fights about Natasha in *War and Peace* will go on as long as they need to. Possible sticking points: Do you think she was truly good? Natasha is selfish in her

attempts to elope with Anatol, but is that attributable to her naïveté? Is Anatol to blame for exploiting her? Is her susceptibility to Anatol's exploitation a character fault or a fatalistic component of youth? Watch sparks fly.

8. If this is your first night at book club, you have to prove that you read. Do so by making mention of small details in the reading, like how on page X the protagonist notes something interesting and uncharacteristic of her typical observations.

Petition to Change the Term from "Bookworm" to "Bookcat"

THE BIGGEST PROBLEM FACING us readers today concerns our mascot, the bookworm. In illustrations it is usually rendered as a chubby brown or green worm with a dopey smile and glasses, sometimes wearing a tie, and either holding a book or eating a hole through it. Readers do not take offense at the idea that we chew through books. We do, however, take affront at the notion of being characterized as worms or cherubic caterpillars. Who decided that the equivalent of a person who reads often is a filthy, writhing grub? What bully exalted the reader's isolationist nature by making them the type of creature no one else wants to be around? Which jock determined that we're ugly and dull instead of something sexier, flashier, like a peacock or a kitten?

In this era of social media and rebranding, surely the same creative geniuses who made Pabst Blue Ribbon go from a cheap untouchable mix of water and grain to the favorite of millions of hipsters can do something about the term that holds back readers most.

The time is now to claim our newest mascot. I propose the cat, for these reasons:

CATS ARE KNOWN FOR BEING LONERS.
Readers enjoy being utterly lost in their own world for hours. A night in is a godsend, not a bore.

CATS DEVELOP QUIRKY HABITS.
You think it's weird when your cat hides between the wall and the back of the couch after it eats? Cats think it's weird that you only read in bed with your right arm tucked under your body and your left hand holding the book up, and you use your nose to turn the page.

CATS TEND TO BE STUBBORN.
Try convincing a self-described "serious" reader to pick up a Harry Potter book. Or a Nicholson Baker fan to settle for a Dan Brown novel. Ever sell a reader on an author they previously despised? Didn't think so.

CATS LIKE TO CURL UP WITH/BY THINGS.
There's a reason the phrase is "curl up with a good book"; we're catlike when we tighten our posture around the book, wrap a blanket around us, and burrow into the most comfortable position possible.

CATS HOLD GRUDGES.

Ever think you'll forgive James Frey for *A Million Little Pieces*? Kind of like how your cat doesn't know if she can forgive your boyfriend for the time he stepped on her.

CATS ARE LOW-MAINTENANCE.

The typical reader needs no sort of accessory other than possibly a pair of glasses and a bit of light.

Fan Letters

HAVE YOU EVER SPOKEN to a Phish fan about their relationship with Phish? It's not a band that they turn on sometimes in the car, whose albums they own all of, or that they go see once a year. It's a life. It's a devotion that goes beyond the music; the fan becomes part of a culture. And this kind of fandom breeds expectation. You can't call yourself a fan unless you fantasize nightly about touring with the band, show up at every concert within a day's drive, and preorder their every release within three hours of its availability. And, take note, the more cultish the band is, the greater the commitment.

Literature fans have it differently. And our fanaticism usually revolves around the books themselves, not the authors. The love is for Ishmael, not Melville. With great authors, it's not until finishing a book that we're reminded there was a puppet master as the author smiles up from the back cover. And even if we do develop crushes on Eugenides or Karr, we can only see our idols if they swing through our city on a book tour.

Not every author has escaped from personal fans. Some authors in particular gather what might be called "cults of personality" around them. They are authors who, because of either the strength of their message, the idiosyncrasies of their personality, or the sheer romance of the worlds they conjure, inspire a devotion akin to the adulation reserved for rock stars and movie stars. But the authors are not the only ones to benefit from this fervor. The fans themselves assume a contact sheen depending on whom they choose to idolize—or they think they do. More often their allegiances reveal their personal aspirations, insecurities, and delusions—all of which probably existed before the idol-author came along. In any case, now allow me to berate the most blatant offenders of those who can be judged by the books they love to read.

Open Letter to Ayn Rand Fans

Oh Christ, we get it. Do you get it? How can you be so focused and not see that you've chosen the most transparent philosophy to live your life by? Women: Do it. Fuck like Francon and fight like Taggart. Go for it. Men: Go. Win. Make money. Don't spend any time at home. Try in vain to create a tortured-genius persona as all-consuming as Howard Roark's. Just don't tell me you actually believe in objectivism as a real philosophy. Rand is an amazing storyteller, but to say that her philosophy is presently relevant is to be willfully blind.

Open Letter to Yann Martel Fans

Fans: let's first give thanks to Moacyr Scliar for handing Martel the fully formed idea for *Life of Pi*. Scliar wrote the tale *Max and the Cats*, about a young man who has run away aboard a ship and ends up on a life raft with a jaguar. You haven't heard of that book? It's basically *Life of Pi* but without the sophomoric writing and overstated messages. And I just have to wonder how you steer around the awkward glances when someone asks your thoughts on Beatrice and Virgil, the genocide allegory that, he stated in an interview, he had the right to write because "the Jews don't own the Holocaust."

Open Letter to Marcel Proust Fans

I have a lot of admiration for you Proust fans. One, you seem to always spit whenever you pronounce his name. This is likely because you're salivating over the pride you have for getting through all seven volumes of *In Search of Lost Time* without realizing how much time you were losing while doing so.

Open Letter to Kurt Vonnegut Fans

You're lazy in your insanity. You're erratic but not spaced out enough to be obscure. You don't care for the norm but you don't care to fall too outside of it. You like to read but you're not a reader. You like to write but you're not a writer. You like to exist but you're not making a statement.

Open Letter to Haruki Murakami Fans

I can say without irony that you like good music. You do. And you know what? You heard about that band first. I get it. You're "in the know." You were the first person ever to hear of the band Phoenix. You were reading Murakami well before legions of college students who thought they were above Vonnegut got turned on to *The Wind-Up Bird Chronicle*. Good thing you understand Murakami better than all those other kids.

Open Letter to Miranda July Fans

It's an unfortunate truth that if everyone is acting "unconventional" with ironic text-speak and quirky, flea market dresses paired with printed tights, they are acting conventional. Although we can never fault July for her brilliance in bringing us the emoticon))>><<(((via her movie *Me and You and Everyone We Know*, we can fault her for the onslaught of slam poet wannabes in our local coffee joints. Sure, she's a welcome substitute for those manic-panic pixie dream girl Zooey Deschanel wannabes, but beyond a cool book title and great social media presence, where's the substance in her writing?

Open Letter to Gary Shteyngart Fans

Do you even know how to spell his name? Or do you know precisely how to spell his name and you correct others on it? Are you only a fan of his because you can correctly spell his name? I bet you purposely type it in very quickly in Gmail Chat, so the person you're talking to about how great Shteyngart is knows that you had no need to Google the spelling beforehand. I'm genuinely curious because the proper choice, if you're trying to be a fan of contemporary satirists, is Sam Lipsyte. So I must assume you've chosen Shteyngart over Lipsyte because (first) you can spell his name and find that impressive and (second) you've been reading too much of the *New York Times*.

Open Letter to Douglas Adams Fans

You throw a party on May 25, Towel Day. You have a DON'T PANIC poster on the wall in your bedroom. You add in "mostly" whenever you describe anything as "harmless." You end e-mails to your friends with "So long and thanks for all the fish." No toast is complete without "Share and enjoy." To you, Adams's *The Hitchhiker's Guide to the Galaxy* is appropriate to quote at any point. Your spastic quirks require you to say, "Forty-two," whenever anyone begins to talk about the meaning of life. Your blog has the tagline "Life, the universe, and everything." Keep in mind you're annoying to those uninitiated in the jargon.

Open Letter to Tim Ferriss fans

Stop it, I don't believe you. I don't believe Ferriss can give a fifteen-minute orgasm to a woman with his *4-Hour Body* as much as I don't believe anyone who picks up *The 4-Hour Workweek* will actually accomplish it. I don't want to see the residue of Ferriss's influence when I e-mail you and receive an automatic response that states that you're busy and will reply if necessary when it's convenient. I want you to suffer alongside me in the dregs of e-mail hell.

The Spelling Bee

THE BEGINNING OF MIDDLE school is when you get classified. Sure, identities can shift and reputations may change, but for most your twelfth year on Earth is what seals your social fate. Especially those suburban kids who have the same classmates K through twelve, as I did. There are the standard categories: dorks, populars, dweebs (weaker than dorks), jocks, punks, etc. The "etc." is a section of circular holes for square pegs, kids who can't fit into any of the groups and instead roll around their ill-fitting section, unable to mark a spot. Members of the etc. are the dirtiest word of adolescence: "unique." Being unique in the classification system is much, much worse than being a part of a larger group. "Unique" is what teachers and parents call the kid who believes himself to be a vampire, the girl who eats her boogers, and the boy who cries in class.

I was an eager kid, eager to show off my intelligence, eager to learn. Being a heavy reader is always a big downfall of the adolescent mental state because you start to expect the experiences that often appear in books. I expected (or even might say wanted, so fully I believed myself to be a part of a grander tale, so thoroughly I invested myself in stories) horrible incidents on my way to adulthood, just like those in the lives of the heroines in my young-adult novels: a first period that bleeds through my white pants in front of the whole lunchroom, mean nicknames about my breasts given to me by my peers, a crush who acts like I don't exist and goes out with the prettiest girl in school, a first kiss couched in bad breath and awkward tongues. I would be wrenched into adulthood by a passageway of mortifying, embarrassing events that would change me. I welcomed it. I knew full well that however bad it got, I'd have a happy, tidy ending just like my admired protagonists. The school would rally around me, the episode would spark insightful conversations, the guy would realize his girlfriend is shallow, and I'd emerge triumphant. For every action, there is a reaction. And those reactions tend to load the victim up with enough good fortune and joy. Unfortunately, life doesn't always mirror fiction. Especially not in middle school.

I was unclassified going into the winter of my sixth-grade year. I was a nothing. I didn't exist. One day, announcements came over the PA that sign-ups for our school's spelling bee would be held in the library after school. I had read enough young-adult novels to know a spelling bee was just the type of rite of passage that a twelve-year-old like me needed to experience. I'd win, the crowd would cheer, my position as a dork and all the bespectacled friends who might come with it would be solidified. I'd gladly take on taunts about my brain size or too-high pants in exchange for being recognized as *something*. At the beginning of the year I might have held hopes of becoming popular, but by four months in I just wanted friends.

I headed to the library and picked up the packet of words they handed out. I snorted at how easy it seemed; the packet was around twenty pages long, but come on! *They gave you all the answers!* I had been expecting to be tasked with reading the whole dictionary, pouring over archaic words like "keitloa" or "*Zwischenspiel*."

For weeks I studied, writing out the words, closing my eyes and spelling them out while on the way to school (I crashed my bike twice), having my mother quiz me. She'd paint her nails, glance at the sheet, and say, "Expeditious."

"E-X-P-E-D-I-T-I-O-U-S."

She'd glance back at the sheet and say, "Languid."

I protested. "Wait! Mom! You didn't check to see if I was right. *You didn't look at the spelling of 'expeditious.'*" A sigh and an eye-roll thrown in my direction. It'd be years until I realized she didn't need to check because she knew how to spell better than I did. I'd grab the sheet from next to her, exposing myself to the correct spelling of "languid" as I verified "expeditious," in the meantime losing memory of how exactly I did spell it. My mother never understood why someone wanted to know *more*. Her daughter's eagerness to prove herself as a lonely dork confused her. I have a feeling that after our spelling sessions she knelt and prayed I'd find an interest in cheerleading (which she had repeatedly signed me up for, despite my lack of coordination and congealment with the fellow cheerers).

My favorite part of the practice was the way you had to spell out the word in competition—"repeat word, spell, repeat word": "Expeditious. E-X-P-E-D-I-T-I-O-U-S. Expeditious." To this day, I have no idea which fork goes where at a place setting or how to properly write a thank-you card, but my etiquette when it comes to spelling a word for someone is pristine. I was in love with the fact that during competition you couldn't go backward, you couldn't say, "Oh wait! I meant I-O-U instead of A-O-U!" Your mind had to nail every letter; it was important to pause and mentally touch every part of the word before speaking.

The day of the event I was jazzed, joyed. Since it was my first year at the school, I hadn't known what a large event they made it. I expected a classroom; I got the auditorium. Each period, all the English class came down to watch, and for the last couple of rounds, the entire school came down. Parents were in the audience as well; my mom eventually showed up, when I was far enough along that it qualified as an "event she should be at" in the parents' handbook.

It was a while before I realized how well I was doing. I was aided by the fact that everyone was watching, flying high on adrenaline. Sometimes, just for effect, I'd ask for a definition. Even better, two boys whom I thought were cute had participated and were already taken out. Surely they'd realize how superior and, thus, immensely eligible I was because I beat them.

It was down to the last three people. I was the youngest kid onstage. Feverish with my good fortune, I started trying to spell faster and faster. "Loquacious. L-O-Q-U-A-C-I-O-U-S. Loquacious." I'd purse my lips afterward, as if I had somewhere better to be and this commitment was wearing me thin. I blew everyone's mind (I was sure of it) when I sped through "Segue. S-E-G-U-E. Segue." I could hear an intake of breath from the audience as they wondered how I could have possibly missed the "W" . . . then the teacher's voice announcing I was correct and everyone breathing out in confusion. I saw my mom smile, relieved I didn't make a rookie mistake on such an obvious spelling bee trap.

People aren't fully formed yet during middle school, they're just globs of hormones and wandering personality traits gained through osmosis from pop culture, with senses of humor consisting almost entirely of canned lines from funny movies. And around that time, Adam Sandler was king. *Billy Madison* was huge, and everyone easily took on his dopey way of losing his temper and cross-eyed mocking.

I was back up at bat. Confident, smiling at my audience. A regular Vanna White with my presentation of letters. The teacher leaned into her microphone and said, "Lauren, your word is 'spaghetti.'"

"Why don't you just *give* her the trophy?!" The line from *Billy Madison*, when he gets frustrated competing against a girl during a spelling bee after she receives an easy word from the teacher, rings out. The entire audience started laughing and I did as well, enthusiastic that everyone was seeing how easily I zipped through the words.

Still laughing along with my audience, I quickly prattled off, "Spaghetti. S-P-A-G-H-E-T-T-Y . . ." And horror struck. The only thing that could have been funnier to my audience than a Billy Madison line was the girl onstage misspelling such an easy word in such a silly way. The audience began hyperventilating with laughter as my face reddened.

Even the teacher seemed tickled as she said, "I'm sorry, Lauren, that's incorrect." Tears filled my eyes. How could I misspell "spaghetti"? After "surreptitious" and "sachet"?

I should've known better than to go to my mom for reassurance after I left the stage. She was still laughing, the same laugh as all the middle schoolers. "My Italian daughter! You eat spaghetti all the time! How do you not know how to spell it?!" Try as I might, I couldn't convince my mom that it was a flub, not an intellectual error, a slip of the tongue due to speed and hubris. My mom kept laughing too loud to hear my protests.

When I went home that night, my mother made spaghetti for dinner. To really drill the point in, she decorated my room with spaghetti boxes, having used a Sharpie to double-underline the "I" on all the boxes. "My Italian daughter! I can't believe it!"

It turned out I couldn't convince anyone at my school as well. No matter how many times I argued that *I knew how to spell "spaghetti," it was just a mistake,* not one person in my life believed it was a slip of the tongue. There was no triumphant ending, only humiliation and a joke that hasn't depreciated but instead gains momentum from its frequent repeating at holidays and bars. I was "the girl who can't spell 'spaghetti.'" The word "spaghetti" in my presence caused snickers for the next six years from my classmates.

It's been well over a decade since that incident and my closest friends who knew me during middle school enjoy making me miserable by bringing up the anecdote in front of new friends, new boyfriends, anyone I introduce to them. The man in my life will tell them, "Lauren makes great pasta sauce." They start to chuckle and ask, "Does she ever make you spaghetti?" I'll palm my face as he nods, privately somewhat impressed with the fervor my friends still feel for this personal joke. "She seemed so smart and then we all found out she couldn't spell 'spaghetti'!"

Your Moveable Feast

I LOVE DINNER PARTIES. Eating too much, laughing with friends, and getting sloshed. None of those is anything less than perfect. I find readers have an affinity for wine foremost—there's something about lingering with a drink that requires interpretation that melds well with our type. Among bookworms there is also a higher-than-average percentage of self-identified "foodies," which might be the most annoying label the *New York Times* has ever propagated. Due to this, mention the word "dinner" and the only event that would get a book lover more excited is if you had said "brunch." Maybe it's the remnants of the Algonquin table on the legacy of being a reader, or maybe we just really love to shovel food in our mouths when we're not devouring words with our eyes, but either way, readers love to eat with friends.

Could you imagine getting the chance to break bread with your favorite author? It's hard, while reading about

Paul Auster and Lydia Davis's marriage, not to wonder what it'd be like to go on a double date with them.

Let's examine what might happen if some famous literary duos were present at your dining table.

Zelda and F. Scott Fitzgerald

F. Scott Fitzgerald, author of the Waspy perennial favorite *The Great Gatsby*, had a tumultuous and rapturous relationship with his wife, Zelda. They were a favorite of gossip circles for their dramatic behavior. Zelda once leapt into a stairwell because Fitzgerald was having an intense discussion with dancer Isadora Duncan. Another time, Zelda collected all the jewelry from guests at a party and—with claims that she was "making soup"—threw all of the jewels into boiling water on a stove. It must be noted Scott was more than just influenced by Zelda: famously, the ending in *This Side of Paradise* is taken straight from Zelda's diary. Scott once recorded Zelda saying she hoped their daughter would be a "beautiful little fool"—the exact same thing Daisy Buchanan says about her daughter.

Imagine drinking with the two of them: Zelda dancing on the table while Scott spills whiskey on your carpet. Zelda trying to throw herself in front of a cab. Scott drunkenly begging after her. God help me if I spend too much time talking to her husband. They'd tell florid stories about their adventures, embellishing the facts and talking too loudly. It would be wonderful. Nothing other than champagne and sidecars would be served and I'd probably get stuck with the bill.

Sylvia Plath and Ted Hughes

Sylvia Plath, poet and author of *The Bell Jar*, was married to fellow esteemed poet Ted Hughes. The couple got hitched only four months after they met and separated two children and six years later when Ted admitted to having had an affair. Five months into their separation, Plath committed suicide by sticking her head in an oven. Six years later, the once-mistress of Hughes who had become his companion murdered herself and their four year-old child by ingesting sleeping pills, turning on the oven, and letting the house fill with gas.

Melodramatic much? At dinner, Hughes will be philandering, Plath will be despairing. She'll take a moment to note the spaghetti you prepared was like ropes of domesticity. Sylvia really shouldn't be drinking with her medication so she'll just sulk. Hughes will be talking too close to my face, spitting about how great their holiday in Spain was and how interesting the people were to watch.

Paul Auster and Lydia Davis

Lydia Davis, revered translator and flash fiction author, and Paul Auster, author of novels heavy with coincidence and failure. Coincidentally, his marriage to Davis failed. This private couple had their relationship exposed a bit by Auster's second wife Siri Hustvedt's roman à clef *What I Loved*. About a year after Auster's son with Davis was convicted as an accessory to a murder, Hustvedt wrote the novel, which involved a middle-aged man with a son going through a very similar-sounding situation while dealing with a steely first wife.

Davis will casually make verbal notations of all the objects around us while avoiding saying anything personal. She'll mention the weather first, in a low voice, and smile in exasperation at the heat. Auster won't sit next to her. We'll have salad and filet mignon, Auster will drink too much red wine, and Davis will give him a look if he indulges in too much red meat. There'll be bitching about an editor and Auster will regale us with funny stories about his neighbors. Davis will grow quieter as Auster grows louder.

Michael Chabon and Ayelet Waldman

Much maligned on blogs for their love of talking about their love for each other, the Pulitzer Prize–winning author of *The Amazing Adventures of Kavalier and Clay*, Michael Chabon, and his wife, *Bad Mother* memoirist Ayelet Waldman, have built a distinct brand around their marital bond. Waldman first raised the ire of the literary press with an essay in the *New York Times* proclaiming her love for her husband over her children and arguing for her right not to apologize for it. The essay also detailed their fantastical sex life, mentioning lube and vibrators.

Heaven forbid a married couple experiment in the bedroom.

The Chabon-Waldman dinner will be full of chatter and opinions—they'll grope each other, overshare sexual quirks, and correct other dinner guests about the best way to tell their kids about the birds and the bees. They're like college freshmen who haven't yet realized it makes friends uncomfortable when he's stroking her inner thigh while carrying on a conversation. But you kind of forgive it, shrug it off. They've been together over a decade; they're the lucky ones. We're all just too jaded to let love be love.

Nicole Krauss and Jonathan Safran Foer

Wundercouple Nicole Krauss and Jonathan Safran Foer found comfort in each other around the same time they found comfort in practically every book critic they encountered. Krauss, author of the novel *Great House* and *The History of Love*, appropriately met *Everything Is Illuminated* novelist Foer while at the Brooklyn Book Festival. Since then, the two have kept details of their relationship private even during a much-speculated-upon episode when the two almost simultaneously released books involving precocious children as their protagonists.

Am I even going to drink? I can possibly imagine crisp white wine coupled with some of the couple's prosaic white whine. Maybe we'll meet for brunch and mimosas. The Brooklyn restaurant where we'll meet has an outdoor garden. They're busy, *they're really busy*, they'll explain when I ask what they've been up to. It's hard to get their kids into the preschool they've chosen. Foer will furtively eye the bacon I'm trying to enjoy and explain the exact process of smoking and curing pork, which I could have already read about in his book, but we both know I haven't bothered to buy a copy.

Vera and Vladimir Nabokov

The Russian-born *Lolita* novelist who described himself as being "as American as April in Arizona" was also a professor, a writer, and a butterfly watcher. Vera and Vladimir Nabokov met while Vladimir was working as a translator in the publishing house of Vera's father. They both had ambitions to be writers, but Vera ended up in Vladimir's service as editor, typist, agent, and driver. Vladimir never learned how to drive. She would type up his novels from the index cards he drafted, leading many to speculate that she had a heavier hand in shaping his stories than she was given credit for. Vera even used to carry a gun to protect her famous and frail husband.

I suspect this couple is not as boring as you might be tempted to assume. Vera will be flitting around Vladimir, making sure he has everything he needs and anticipating any requests before he can even speak them. Vera will be consistently self-effacing; I'll try to compliment her on her dress or shoes and she'll say, "This old thing?" Dinner will be a hefty Italian meal with red wine. Vladimir will have a beer.

I'll ask Vladimir a pointed question about one of his books, why he chose this word or another. He won't remember exactly. Vera will pretend to ignore the question. Another dinner guest will later whisper in my ear that "Vera probably chose that word." Vladimir will make us all laugh with a story about how his students continue taking notes even if he's repeating a paragraph he just read.

Kathryn Chetkovich and Jonathan Franzen

Author Kathryn Chetkovich, longtime girlfriend of the oft-worshipped Jonathan Franzen, once wrote an essay about the frustration of being in the same profession as her immensely accomplished significant other. That essay, "Envy," unfortunately remains her best-known work. The story goes that they met while they were both struggling writers; Franzen eventually churned out *The Corrections* and was blasted to the literary forefront, whereas Chetkovich still hadn't found her voice. The result, as she makes clear in her essay, was unadulterated envy.

I'll endure Chetkovich's glare when I compliment Franzen on *Freedom* but get on her good side by mentioning that I once read one of her short stories somewhere (I made sure to Google it before they arrived). We'll meet for dinner at a dark bistro in TriBeCa. Chetkovich won't be able to make up her mind and will have everyone else order before her. She'll end up picking the lamb.

Little-Known Treasures

PEOPLE TEND TO FOLLOW the pack with their taste in books, abiding by the tides of popular opinion to figure out the next read on their list. Just look at Oprah's Book Club. Even those who try to read outside the mainstream get caught up in pockets of influence; William Gibson's cyberpunk begets Orson Scott Card's space epic begets Susanna Clarke's magical alternative history. But where is the fun in reading the same thing everyone else is reading? The best books are the ones you hardly hear about. These are a couple of cult favorites for people who want to tread the unbeaten paths of contemporary literature.

THE POSTMAN ALWAYS RINGS TWICE BY JAMES M. CAIN
A great, grimy, gritty crime novel to read on an airplane instead of Stieg Larsson or Harlan Coben.

THE DUD AVOCADO BY ELAINE DUNDY
Playful, flaky chickadee romps around Paris. Read this instead of *Breakfast at Tiffany's*.

THE HOUSE OF THE SPIRITS BY ISABEL ALLENDE

Magical realism fans have probably already read this. For everyone else, it's a better introduction to the genre than *A Hundred Years of Solitude*.

THE MASTER AND MARGARITA BY MIKHAIL BULGAKOV

If you're already into Russian literature, it's only a matter of time before you discover this work. The influence by Tolstoy and Dostoyevsky is obvious in parts and yet Bulgakov is utterly original. Ignore *Poor Folk* for this book.

THE MEZZANINE BY NICHOLSON BAKER

A trip up the escalator comes to mean so much more when Baker explicates every thought that runs through the protagonist's head. Replace *Consider the Lobster* with this.

THE ELEGANCE OF THE HEDGEHOG BY MURIEL BARBERY

A remix of the classic class-consciousness tale, read this instead of *Water for Elephants*.

THE LOSER BY THOMAS BERNHARD

One of the denser reads you'll ever encounter, Bernhard makes sure you work for his meanings. Cross off *Portrait of the Artist as a Young Man* by James Joyce in favor of this piece.

THE GOOD SOLDIER SVEJK: AND HIS FORTUNES IN THE WORLD WAR
BY JAROSLAV HASEK

This unfinished novel makes hilarity out of war. Joseph Heller is quoted as saying that if he never read Svejk he wouldn't have written *Catch-22*. Replace the solemn *All Quiet on the Western Front* with this book.

ENDER'S GAME BY ORSON SCOTT CARD

A cult favorite of techies, a young kid is trained to save the world. Much better than *Rama II*, I promise. If you read this and like it, check out *Snow Crash* by Neal Stephenson.

SHARDS OF MEMORY BY RUTH PRAWER JHABVALA

Seemingly simple tale of a family held together by a cult reveals the complexities of familial bonds and hierarchies. Read this instead of *Gilead*.

THE SECRET AGENT BY JOSEPH CONRAD

Put this book before any other Conrad work. *The Secret Agent* is part spy thriller, part meditation on the life of one aging spy.

THE LIFE AND OPINIONS OF TRISTRAM SHANDY, GENTLEMAN
BY LAURENCE STERNE

Credited as one of the first postmodern novels; read this instead of anything by Jonathan Safran Foer.

Infinite Lies

AFTER DAVID FOSTER WALLACE died, a friend instant-messaged me for my reaction. I had seen the articles and digested the obituary; this was a couple days afterward. I expressed the appropriate sentiments to him.

"Have you read *Infinite Jest*?" he asked.

"Yeah, loved it," I replied.

Except I hadn't.

I surprised myself. I'd never lied so baldly about having read a book that I hadn't. Sure, I'd massage the truth sometimes, mentioning bits without mentioning that I'd gathered them from a review or skimming the back cover. But never before had I directly lied.

I was bitter about the whole thing, mad at myself for not realizing it was risky to not be in any hurry to read his books. Philip Roth, John Updike, Don DeLillo—there was a long list of contemporary authors who could've suddenly dropped dead and we would've been sad, sure, but not caught off guard. I was slow to approach contemporary literature, budgeting my high school years to creep through the postmodernists. I ignored the "New Releases" table at the bookseller, thumbing my nose at hardcovers. I had no time to waste on literature's youth, passing on opportunities I had to attend lectures in favor of curling up with a long-dead author. I figured I'd arrive at the present, contentiously framed post-postmodernism era around the time I'd graduate college, with plenty of time to spare for visiting authors' signings and listening to readings. I planned on moving to a big city, where writers like Wallace would be sitting in the coffee shop around the corner; that's when it would be important to have read his book. However, I hadn't planned on any of those young writers disappearing.

So I said, "Yeah, loved it," because I should've read it. I should've read it and shown up at nearby schools when Wallace visited to talk about his work. What better experience could there be than asking an author about their work? Why was I so opposed to acknowledging the living, hell-bent on studying only those already passed or about to croak?

The time-and-tragedy combination increases notoriety. Wallace called himself "as famous as your local weatherman" but made it to national-news-anchor-level fame after passing. Compare this effect to the death of another nineties meteor, Kurt Cobain; absence makes fans' hearts grow fonder. David Lipsky got a book deal for interviews he conducted with Wallace—previously deemed too uninteresting to publish by *Rolling Stone*. The resulting book is now a *New York Times* bestseller. A fan site for the author, The Howling Fantods!, saw its traffic double in the months leading up to the publication of Wallace's posthumous *The Pale King*. Unfortunately for me, the friend whom I bluffed about reading *Infinite Jest* to hasn't stopped mentioning the book every time we hang out, which thankfully has been less and less over the years. His insistence on bringing up the book has me convinced he's trying to trip me up; he could see through me over instant message and wants to punish me for my lie. Luckily, for all his talk of the book, he still hasn't tried to read it, so I'm safe from grilling on specifics. Our friendship will officially end on the day he creases the spine.

I owned the book by the time I moved out to New York, somewhere along the way acquiring it, knowing now was the time to own it, even if there was no chance of turning a corner and bumping into Wallace. New York had everything it had advertised: readings, signings, book clubs. I decided to join a reading group at the Center for Fiction (a *center* for *fiction*, can you imagine how happy that made me?) for *Infinite Jest*. The cost of the class was $80, a price I felt was a significant burden that would nail me into attending the nine meetings they had, once a month.

"Have you read the book before?" a coworker asked me after I announced my plan at lunch.

"Yeah!" For the life of me, I have no idea why I lied again. Continuity? Self-flagellation?

"Cool, care if I sign up to take the class with you?"

"Sure, have you read the book before?" I asked, not wanting to hear her answer.

"Yep!" So now I had lied to her, and she'd be accompanying me to every class, expecting me to not be surprised by plot twists or unsure of the direction in which a character was developing. She then told me about her experience reading the book: she had read it during college with a group of friends, ripping through the book over two weeks of marathon reading sessions and heady discussions. When she finished, she looked at me as if it was time for my tale of reading *Infinite Jest*. I smiled and walked quickly back to my desk. What part of the fact that this was one of the most well-known and talked-about books in the last twenty years did I not understand? Other, dutiful and honest people have read it. The book was my telltale heart; everyone I lied to seemed to be suspicious of me. She knew, now that I had supplied no story for how or why I read the book, that I hadn't. How does one get through an over-nine-hundred-page book without some sort of event, an anecdote to color the experience?

Another friend signed up for the class with me as well; I hadn't lied to her when she asked, "Have you read the book?" I'd replied, "Bits and pieces," which was a significantly more truthful version of my lie, or at least I thought so.

On the day of the first class, my two friends and I sat together. The man who led the Wallace reading group was a mellow, middle-aged academic who said things such as "I'm a Joycean by nature" to apologize for future gaps in his postmodern knowledge. Something about him reminded me of Bob Ross from PBS, softly guiding us on how to dab the paint to make a bird against a sunset. "Now, let's go around the room and everyone should tell us their name, if they've read *Infinite Jest* before, and what they're hoping to get out of this class," he told the group.

I'm an idiot, I thought. I hadn't realized that telling different lies to the two people I was going to be sitting in between in the group discussions would create a minefield. I cursed the teacher, pathologically assuming this was his way of outing liars. I'd have to say yes, and then my experience in the class would be marked by the fact that supposedly I'd already read the book. It wouldn't just be my coworker I'd have to perform for, it'd be the entire group. Even worse, I had to find a middle ground between having read "bits and pieces" and having read the whole book to answer his prompt of whether or not I had read it.

"I'm Lauren; I've read . . . *most* of *Infinite Jest* . . . [lifting from my coworker's experience] I read it in college, not as thoroughly as I'd like [ha!], so I want to return and read it well." Read it well, indeed. I couldn't look at either friend afterward, hoping to avoid questioning.

When it came to equipment for reading *Infinite Jest*, I spared no expense. I downloaded the book on my iPhone, brought the print version to work with me, got an electronic copy for my iPad at home. I was ready to conquer the thing, to save myself from my lies by eradicating the issue.

I made it to the second class of the discussion group, then

I never made an appearance again. Time was an issue. It was also frustrating to have to start and stop, to avoid getting too far ahead or behind during the month we had in between classes. Discussion groups generally turn out to be more of a burden than a blessing; twenty people on one book can't be efficient. But most frustrating of all was reading Wallace, knowing that you'll never hit on his level, that you didn't even know of his level when you arrived. He made me throw down his book, try to write something myself, then study it for possible similarities or traces suggesting that I could do the same. They didn't exist. He was the best thing for my writing and the worst, forcing me to try again but also lying out of reach and causing anxiety over whether I ever could write well enough. I found it slow going, being forced to actually read in bits and pieces (a bit of cosmic retribution) because of the sugar-high-like feeling I'd experience while being confronted by page after page of good writing. A friend of mine expressed once that the closer he gets to the end of a book, the slower he moves. He doesn't want it to finish, he says. I never understood that feeling until I read *Infinite Jest*. You just want it to go on.

It's telling that the teacher of the class labeled himself a "Joycean." I've found that James Joyce, for much of the generation of writers preceding Wallace, is the most-often-mentioned name when they discuss influences. I find Toni Morrison noting to *The Paris Review* that she appreciated Joyce's humor and irony. Norman Rush described Joyce as a "wondrous and calamitous influence" on him. Salman Rushdie claimed, "Joyce is always in my mind, I carry him everywhere with me." There's some inexplicable connection to understanding how or why an author writes the way they do and what they thought of *Ulysses*. In the last five or so years, Joyce has been replaced with Wallace by authors explaining influences. Neal Stephenson said, "I think [he] was the best we had, and who influenced me in the sense of making me try harder and wanting to do better." Teddy Wayne spoke about how he wrote his English thesis on Wallace. Dave Eggers described his writing as "world-changing." Zadie Smith stated, "I didn't feel he had an equal amongst living writers." This is the greatest point in any argument I can make when I try to claim Wallace's style was profound enough to hurtle us into a new realm of fiction; there have been many, many other writers who have jotted down things and examined them with hopes the scribblings might miraculously contain some Wallace elements.

I am happy to say, I now have finished the book.

There, you see? I lied again.

How to Write Like Any Author

WHEN YOU'RE YOUNGER, your writing takes on the cadence and tones of your newest favorite author. Suddenly, during your J. D. Salinger phase (which is when most of us start to write, unfortunately), you're sarcastic and skeptical about the booze in your cup at a party. As you move on to Hemingway, you're brief and unmoving about otherwise complicated emotions and situations. In college you begin to use more flowery descriptions; you're reading the Russian greats and suddenly the manuscript you've been working on gives mention of a bloodline in an otherwise contemporary family.

Then, one day, you stick your own voice in there. And it's amazing, because you didn't know you were missing originality; you didn't even realize how heavy-handed you were with your inspiration from the greats. Your warbled, watered-down versions of their voices have now, with the addition of your own self to your writing, become *influences*.

For the time when you're still trying to find your way on a page, here are some tips for writing like others.

STIEG LARSSON

Your male lead should have an inflated sense of self. Your female lead should have a deflated sense of self. The only props any character may be fiddling with are drinks, coffee, cigarettes, and weapons.

JOAN DIDION

Be redundant, be scattered. Try not to stay on the same point for any real detail. Find beauty in simple objects. Have little epiphanies in every paragraph.

SIMON RICH

Your character's inner dialogue should read like a cross between the thoughts of a neurotic sixty-year-old Jewish man and those of a precocious seven-year-old. Do not bother spending time on character development; instead focus on a half-dozen supremely well-crafted observations.

MALCOLM GLADWELL

Blame all innovation and talent on blind luck. Give nothing more than anecdotal evidence for every thesis you put forth.

JHUMPA LAHIRI

Write about a husband and wife who are finding excitement or strife in their daily life. The writing should be perfect yet plain. Your MFA professor will be proud.

HENRY MILLER
Sentences cannot go beyond a line and a half. Preferably they should be under ten words. If a sentence must go on, make sure it is laden with em dashes.

ANAÏS NIN
Find a way to make the most graphic visuals ever without actually mentioning the proper names for genitals.

DON DELILLO
State the object. Then, in the next sentence, give an adjective to the object. Describe the object now with scientific detail about its appearance. State the object again.

CORMAC MCCARTHY
Every paragraph must have a sentence fragment that somehow, magically, conveys a whole thought with only three words and rarely a vowel. Bonus points if you can accomplish it with two words.

MARILYNNE ROBINSON
You write, and they read, and you write more, about everything and about anything else in there and all that. And then you deny editing; instead you check it again, because you don't want to conform and you're afraid of seeming like everyone else, you tell me.

ANN BEATTIE

It's this writing that lets you go on and on, expanding the story line, extensively delving into the actions of the protagonist, and so on, simply by adding a comma, you see? So similar to Raymond Carver, in her style of short stories, but with so much more punctuation, it's really quite interesting.

RAYMOND CARVER

First-person exploration. Beautiful woman talk to me, No quotation marks necessary.

JUNOT DIAZ

Pop culture reference, Spanish phrase.

PART III

How to Fake It

What Your Child Will Grow Up to Be If You Read Them . . .

MUCH LIKE ADULTS ADOPT favorite authors and genres in a way that defines their tastes, children fall in love with a particular imaginary world that embodies the elements they hold dear. For many, that's their first bit of rebellion, recognizing that not every book is on the same moral plane when you laugh much harder at Dr. Seuss's ne'er-do-well cat than the goody-two-shoes big red dog named Clifford.

A parent chooses (consciously or otherwise) books in line with the message of their parenting style; something in the plot picks up on how they'd rather their child appear, as the subdued orphan Madeline or the wily, brash Eloise. Remember what your parents read to you when you were younger? Did they try to quietly influence your worldview by stockpiling your tiny bookshelf with *Paddington Bear* instead of *Tom Thumb*? Now you have more evidence than their divorce to mount on the wall of things they did wrong.

For the record, my mom read *Love You Forever* to me so much as a kid that I can repeat it by memory, but my favorite books were the Berenstain Bears collection. I was an annoying child.

THE GIVING TREE

"I want that!" your daughter or son will be yelling next time you take them to the mall. Smooth move, Mom and Dad, teaching your kid about how parents will sacrifice for their children. Have fun explaining to your kid why a tree would do more for him than you would.

CLIFFORD

A big, red dog who . . . wait, does Clifford ever do anything? It's a boring book for boring kids. A gigantic dog that doesn't eat humans or at least crush a couple of houses? Snooze fest. Congratulations on begetting an average kid who will find a great position in middle management and have a somewhat happy marriage.

GREEN EGGS AND HAM

The perennial favorite of any gross-out kid—the one who throws boogers at classmates, pees his pants laughing at fart jokes, and pretends to blow up nearby buildings with the air bazooka he's holding. In short: *awesome*.

THE VELVETEEN RABBIT

A crybaby in the purest sense of the term. Your child will

get sentimental and emotionally attached to seemingly any object. These are the kids who won't let a balloon go until it's wilted on their bedroom floor. They'll try to keep June bugs as pets. If they went to Hogwarts, they'd get placed in Hufflepuff.

LOVE YOU FOREVER

Daddy issues. The entire book makes not one mention of a father. Also, the whole taking-care-of-your-mother-as-an-old-woman thing is deeply creepy. Five-year-olds don't need to know about that.

A LIGHT IN THE ATTIC

The edgy alternative to Shel Silverstein's *The Giving Tree*. Parents, you're going to end up with a stoner on your hands. A wonderful, happy, creative pothead. How could you not? This collection of Silverstein poems is best read high. It's only a matter of years before your kid tokes up and pulls out this book to blow his mind.

THE LITTLE PRINCE

Quietly contemplative. Regal. The other kids might call him names but you know your child is just introverted. He is far too busy contemplating the tao of life to bother with child-hood rowdiness.

HAROLD AND THE PURPLE CRAYON

When your grown-up kid is delivering the news of his second

or third divorce, remember all the nights you spent reading him this book, telling him that a boy could dance around with a purple crayon and make anything he wanted come to reality. Then stop wondering why he won't stop sleeping with his secretaries. The answer is right here.

CURIOUS GEORGE

The creepy monkey kid whom no one likes. Seriously, why did the kid who loved monkeys always also look like a monkey? If you pick your kid up from school and a bunch of the other students are having him imitate an ape, you have to put your foot down. If not, in about ten years no one will go to prom with monkey kid.

WHERE THE WILD THINGS ARE

Navel-gazing Tumblr addict. Seriously, Spike Jonze makes a movie about a children's book and suddenly every hipster "omg < 3"s it. Next.

ELOISE

"Narcissist" is too easy and much too simple a word to describe Eloise fans. Those nosy gossips with a taste for high-class clubbing and the ability to seek out the best sample sales will be moving straight to a big city after college graduation. Just wait for their e-mails eagerly sharing their photos on The Sartorialist and mentions in the local gossip rag.

THE BERENSTAIN BEARS

Wild, hyperactive kid who gets kicked out of class for laughing too hard at things that aren't funny. They're unable to gauge when and how to end the joking around. If you oblige them with a chuckle, you're inviting them to beat the joke into the ground.

THE WIND IN THE WILLOWS

Boring, crusty-nosed girl with glasses who hangs out in the library. I'm yawning just thinking about that book.

PANCAKES FOR BREAKFAST

This is a great book to read to your child if you never want to see your child again. You're creating an adventurer. The "I'm going to move to Denver for a couple years and see how it pans out instead of graduating college!" kid. The "I think it's a great idea to move to Paris without a job and see where life takes me!" kid. It's awesome that you've created an independent adult but you might get lonely on the holidays.

MADELINE

Horrifyingly obedient, to the point where you can be reassured that even if you traveled out of town for a month and left her alone, your teenage daughter wouldn't dare throw a party or look twice at the liquor cabinet. As an adult, she becomes a church group leader, even though you didn't raise her with religion.

Stereotyping People by Favorite Author

THERE'S SOMETHING ESSENTIAL ABOUT our choice of our favorite author or book. We love those who speak to our experiences or to what we wish our world to be like. The author we put on a mantel and formally designate as our favorite says something very real and fundamental about how we view the world. A science fiction fan loves the alternate reality a book presents but knows it's not real. From those dystopian societies, she doesn't derive the expectation that she'll someday live on Mars—she takes, instead, the feeling of adventure and endless possibilities. A Faulkner fan might not live in the South nor have any plans to, but he believes in the power of families and small communities. Your favorite author frames your approach to life. Having our favorite authors' names proudly displayed on our bookshelves is our way of most aptly expressing the otherwise inexpressible.

I am hardly the first person to point out that a particular

author's fans often share a distinct personality type. Martin Amis once said of his experience at a book signing with Roald Dahl, "[At] signing sessions with other writers . . . you look at the queues at each table and you can see definite human types gathering there." Like attracts like—and here are some authors who seem to attract a particular breed.

J. D. SALINGER
Kids who don't fit in (duh).

STEPHENIE MEYER
People who type like this: "OMG. Mah fAvvv <3 <3."

J. K. ROWLING
Smart geeks.

JACK KEROUAC
Umphrey's McGee fans.

JEFFREY EUGENIDES
Girls who didn't get enough drama when they were younger.

LAUREN WEISBERGER
Girls who can't read. Or think.

JONATHAN SAFRAN FOER
Thirtysomethings who were cool when they were twenty-something.

JODI PICOULT
Your mom when she's at her time of the month.

CHUCK KLOSTERMAN
Boys who don't read.

CHUCK PALAHNIUK
Boys who can't read.

CHRISTOPHER HITCHENS
People I would love to hang out with.

LEO TOLSTOY
Guys I want to date.

FYODOR DOSTOYEVSKY
Guys I want to sleep with. (The difference between Dostoyevsky and Tolstoy lies in the fact that I think the Underground Man is sexier than Pierre Bezukhov.)

CHRISTOPHER BUCKLEY (OR WILLIAM F. BUCKLEY)
People who love excess verbiage.

AYN RAND
Workaholics seeking validation.

DAVID FOSTER WALLACE
Confirmed nineties literati.

JANE AUSTEN (OR THE BRONTË SISTERS)
Girls who made out with other girls in college when they were going through a "phase."

HARUKI MURAKAMI
People who like good music.

RALPH WALDO EMERSON
People who can start a fire.

NATHANIEL HAWTHORNE
People who used to sleep so deeply that they would pee their pants.

CHARLES DICKENS
Ninth graders who think they're going to be authors some-day but end up in marketing.

WILLIAM SHAKESPEARE
People who like bondage.

MARK TWAIN
Liars.

SIR ARTHUR CONAN DOYLE
People who drink scotch.

JOSEPH CONRAD

People who drink old-fashioneds.

DOMINICK DUNNE

People who get their class from *Vanity Fair.*

ANNE RICE

People who don't use conditioner in their hair.

EDGAR ALLAN POE

Men who live in their mother's basements. Or goth seventh graders.

MICHAEL CRICHTON

Doctors who went to third-tier medical schools.

DAN BROWN

People who used to get lost in supermarkets when they were kids.

DAVE EGGERS

Guys who are in the third-coolest frat of a private college.

EMILY GIFFIN

Women who give their boyfriends marriage ultimatums.

RICHARD RUSSO
People whose favorite day in elementary school was Grandparents' Day.

ANAÏS NIN
Librarians.

MARGARET ATWOOD
Women whose favorite color is hunter green.

WILLIAM FAULKNER
People who are good at crosswords.

JACKIE COLLINS
Your drunk stepmother.

NICHOLAS SPARKS
Women who are usually constipated.

JAMES PATTERSON
Men who score a 153 on their LSAT exam.

SYLVIA PLATH
Girls who keep journals (too easy).

GEORGE ORWELL
Conspiracy theorists (too easy).

ALDOUS HUXLEY
People who are bigger conspiracy theorists than Orwell fans.

HARPER LEE
People who have read only one book in their life, *To Kill a Mockingbird* (their assigned reading in the ninth grade, naturally).

NICK HORNBY
Guys who wear skinny jeans and the girls who love them.

ERNEST HEMINGWAY
Men who own cottages.

F. SCOTT FITZGERALD
People who get adjustable-rate mortgages.

VLADIMIR NABOKOV
Men who use words like "dubious" and "tenacity."

FRIEDRICH NIETZSCHE
Sommeliers.

BRET EASTON ELLIS
Foo Fighters fans.

HUNTER S. THOMPSON
That kid in your philosophy class with the stupid tattoo.

CORMAC MCCARTHY
Men who don't eat cream cheese.

THOMAS AQUINAS
Premature ejaculators.

PEARL S. BUCK
Women whose favorite president was Harry S. Truman.

THOMAS PYNCHON
People who used to be fans of J. D. Salinger.

ELIZABETH GILBERT
Women who liked the movie *Divine Secrets of the Ya-Ya Sisterhood* but didn't read the book.

REBECCA WELLS
Women on the East Coast who wish they were from the South.

TAMA JANOWITZ
Cougars who went to an urban college in the eighties.

ALICE SEBOLD
People who liked *Gilmore Girls*—even the first season.

MICHAEL SWANWICK
Men who argue Neil Gaiman is overrated.

TERRY GOODKIND
People who have never been Dungeon Master but still play D & D.

STEPHEN KING
Eleventh graders who peed their pants while watching the movie *It*.

H. P. LOVECRAFT
People who can quote Comic Book Guy from *The Simpsons*.

THE BROTHERS GRIMM
Only children with Oedipal complexes.

LEWIS CARROLL
People who move to Thailand after high school for the drug scene.

C. S. LEWIS
Youth group leaders who picked their nose in the fourth grade.

ELMORE LEONARD
People who know how to perform a "Michigan left."

SHEL SILVERSTEIN
Girls who can't spell *"l'chaim."*

DOUGLAS ADAMS
People who bought the first-generation Amazon Kindle.

TUCKER MAX
Guys who haven't convinced their girlfriends to try anal yet.

ALEXIS DE TOCQUEVILLE
Political theory and constitutional democracy majors.

TOM CLANCY
People who skipped gym by hiding under the bleachers.

HERMANN HESSE
People who own one straw chair in their house.

PHILIPPA GREGORY
Women who have repressed their desire to go to Renaissance festivals.

GABRIEL GARCÍA MÁRQUEZ
Men who can't lie but will instead be silent if they know you don't want to hear the truth.

SUSAN WIGGS
Older women who are surprisingly loud during sex.

NICOLE KRAUSS
Girls who intern at *Nylon* but end up moving back to the Midwest for their real job.

MITCH ALBOM

People who didn't go to college but do well on crossword puzzles.

STIEG LARSSON

Girls who are too frightened to go skydiving.

SUE GRAFTON

Women who have an @aol.com e-mail address.

SETH GRAHAME-SMITH

People who own a smart phone that requires a stylus to use.

DAVID BALDACCI

No one. Even the police prefer Tom Clancy over Baldacci.

MICHAEL POLLAN

The girl who just turned vegan to cover up her eating disorder.

ANDREW ROSS SORKIN

People who refer to themselves as "playing devil's advocate."

O. HENRY

Men who have names like Earl or Cliff and were really close with their paternal grandfather.

MICHAEL CHABON

People who hate Ayelet Waldman.

RAY BRADBURY
People who own souvenir golf club covers.

JOSEPH HELLER
People who love buying drinks for their friends. Also, people who cringe when they read their bar tab.

DAVID MITCHELL
Women who live in any area of Brooklyn other than Park Slope but may end up there someday (and, if that day comes, they will switch to Barbara Kingsolver).

MAX BARRY
People who don't mind the color orange.

DEAN KOONTZ
People who would never dream of owning any type of "toy"-breed dog.

JOHN IRVING
People whose parents are divorced.

RICHARD DAWKINS
People whose significant others grab them under the table to shut them up whenever a dinner guest says something absolutely ridiculous and wrong.

SALMAN RUSHDIE
People who Google-image-search Padma Lakshmi late at night.

ALBERT CAMUS
People who went to art school after "trying it out" at a public university.

KURT VONNEGUT
People who play "Creep" by Radiohead while having sex or smoking pot.

JAMES JOYCE
People who do not like John Cusack movies.

CHARLAINE HARRIS
Elementary school teacher's aides.

JORGE LUIS BORGES
People who took care of their dying grandparents.

TERRY PRATCHETT
People who really like monkeys.

OSCAR WILDE
People who can't resist anything. See also people who claim they're going to change but never do.

TRUMAN CAPOTE
People who would never dream of owning anything that could be classified as a knickknack.

TOM WOLFE
People who don't mind others smoking around them.

NEIL GAIMAN
People who can name at least two Miyazaki films.

J. M. COETZEE
People who correct you on the pronunciation of words like "bruschetta" and "pied-à-terre" but cannot speak a language other than their native one.

KEN KESEY
Wavy Gravy.

TOM ROBBINS
Über-fans of the Butthole Surfers.

THOMAS MANN
Men who don't understand the irony behind hipsters' mustaches.

PRIMO LEVI
Your European Intellectual History professor.

SAMUEL BECKETT

People who loved *Franny and Zooey* by J. D. Salinger.

MIRANDA JULY

Girls who want to be described as "quirky."

HENRY MILLER

People who like smut. Also, people who use the word "smut."

JONATHAN LETHEM

People who secretly love John Grisham.

TEDDY WAYNE

People who just decided to start reading edgier contemporary fiction.

NICHOLSON BAKER

People who get into some kinky shit

PER PETTERSON

The strange man who can't resist telling you "how great that issue" of McSweeney's is when you pick it up at the bookstore.

CANDACE BUSHNELL

Women who think fraternity and sorority life matters in college.

PAULO COELHO
A free-spirited PC user.

YANN MARTEL
Your annoying neighbor.

JOSHUA FERRIS
Someone who hasn't read *The Unnamed*.

AUGUSTEN BURROUGHS
The caterer at the last soiree you attended.

KURT ANDERSEN
People who are sick of hearing parallels between him and Nick Denton.

ADAM HASLETT
The most boring guy at the dinner party.

JONATHAN DEE
Someone who has no idea what memes are and has no interest in finding out.

How to Fake It

WHAT FOLLOWS IS EVERYTHING you need to know to casually discuss some of the most well-known classic and contemporary authors. Pair these profiles with my section on how to avert discussion of major plot points in a novel and you can hold your own anywhere, under any circumstances.

I've broken each author down into three easily digestible sections:

Basics: The bare-bones facts about the author. The author's genre, dominant themes in their work, and bits about their life story. Where it bears mentioning, I have also included the author's influences and whom they influenced.

Essentialography: The three most famous or infamous works by the author that you should know, arranged chronologically. I give a sentence-long paraphrase, then five words to add upon the narrative, a sort of slogan or mnemonic device. Since I tried to keep the synopsis as short as possible, please forgive me if I left out a character or part of the plot from a beloved book.

Details: Nitty-gritty and titillating bits of the author's biography, legacy, and body of work, alongside advice on how to deliver these gems in conversation to a rapt audience.

How to Fake Like You've Read Dostoyevsky

BASICS

This classic Russian author's novels and stories revolve around the character of the "positively beautiful individual"—a person who gives more to others than to himself—and the damage that can result from an individual's purposeful alienation from society and subsequent selfishness. Think of his theme as the opposite of Charlie Sheen's "winning" campaign. (There is nothing sexier than a man or woman who believes they should give more than they get, so keep that in mind next time your date mentions their Russian literature degree. For more advice in this department see "Ten Rules for Bookstore Hookups," page 25.)

ESSENTIALOGRAPHY

Notes from Underground

A Dostoyevsky novella considered one of the first existentialist literary works. The novella's narrator, the so-called Underground Man, set the standard for antisocial behavior and antinihilistic sentiments.

Five words: "Intense pleasures occur in despair."

Crime and Punishment
Dostoyevsky's most popular work during his lifetime. In this novel a man commits a double murder out of folly and wrestles overpowering feelings of guilt in the aftermath.
Five words: Guilt ruins best-laid plans.

The Brothers Karamazov
Dostoyevsky envisioned this novel as the first part of a series that would be his magnum opus but died months after its publication. In it, a dastardly father is murdered by his illegitimate, disabled son and his legitimate son stands wrongly accused.
Five words: Bad girls (Grushenka) are hot.

DETAILS

1. Knowing Dostoyevsky's first name is not important. Knowing how to pronounce his last name *is* important. Say it like this: Dos (like the "dos" in "Dos Equis") toy (the dirty kind) EV (the nickname for that Russian guy you slept with) ski (the sport I cannot do).

2. Note: Just because Tolstoy and Dostoyevsky are both Russian novelists does not mean that you should put them in the same camp. The fact that both wrote long books with characters named Alexey or Alexei and Sophia or Sofia does not mean they're spouting the same message. Tolstoy focused on rationalism, whereas Dostoyevsky thought faith was the only path to

salvation and sought to prove how rationalism leads to nihilism and thus a populace insensitive to beliefs other than their own.

3. Simply put, like *The Tell-Tale Heart* (a play you saw in tenth grade), *Crime and Punishment* is about the corrosive power of guilt; after committing murder, a man's sense of guilt eats away at him psychologically (very bare bones of the plot—try not to talk specifics). No one who is a fan of Dostoyevsky would only read *Crime and Punishment*. Name-drop *The Brothers Karamazov*; say that you still read over Zosima's passages when you want to feel inspired.

4. Dostoyevsky was exiled to Siberia by the czar because he's a badass motherfucker. How badass? Take this story into account. The Russian government sentenced Dostoyevsky to death for being a member of a liberal, intellectual group called the Petrashevsky Circle. Dressed in the outfit of the condemned—a peasant shirt and hood—he was tied to a stake alongside two other men before the squad. Dostoyevsky then heard drums signifying that he and the others were pardoned from death. Instead he was sentenced to hard labor in Siberia for four years. Anyone who can live through that and have any sort of restraint to write novels (the other two men went insane from the trauma) must be a master of Zen.

5. Dostoyevsky focused on ethical questions such as the role of free will and God in a man's life. He wanted people to understand that depravity breeds depravity. Hopelessness will lead to hopelessness. Basically, if you act like an antisocial asshole, you'll be an antisocial asshole (see *Notes from the Underground*). This is because Dostoyevsky was against nihilism and rationalism. He wanted to show how thinking divorced from emotions leads to actions divorced from emotions.

How to Fake Like You've Read David Sedaris

BASICS
Humorous memoirist and essayist who is outshined on NPR only by Ira Glass.

ESSENTIALOGRAPHY
Naked
Sedaris's collection of essays recounting his adventures growing up in Raleigh, North Carolina, and family life under an endearingly jaded matriarch.
Five words: Relatable anecdotes of adolescent embarrassment.

Dress Your Family in Corduroy and Denim
Heartening, hilarious tales of heartbreaking experiences.
Five words: Must read "Full House," "Hejira."

When You Are Engulfed in Flames
Revealing preoccupations with death and inspired ideas about love.
Five words: Quitting smoking and showing qualms.

DETAILS

1. There can be no conversation about David Sedaris without mention of his equally witty sister, Amy Sedaris. You read and loved her parodies of homemaking guides, *I Like You* and *Simpler Times*. *I Like You* is her odd and zany do-it-yourself guide. Your friend actually made Amy Sedaris's pantyhose-and-beans eye burrito one hungover Sunday! You bought *Simpler Times* for your mom on her birthday. Most notably perhaps, Amy was the lead actress in *Strangers with Candy*. Read David Sedaris's *Me Talk Pretty One Day* if you want to hear the best stories about Amy.

2. Spend at least two minutes fawning over Hugh. Hugh is Sedaris's boyfriend, who shows up often in his books. Say you fell in love with him during *When You Are Engulfed in Flames*. I don't care if you're a straight guy; you can't help but love Hugh.

3. Mention you always remove books and magazines from sight when you have company since you read this statement of Sedaris's in the *New York Times*: "If you leave them on the table, it looks like you set them out

on purpose. . . . It looks so phony." Ignore the irony of
the fact that you found that advice in this book.

4. Sedaris has admitted that bits and pieces of his stories
   might be conflated for humor's sake. If the person
   you're engaged in conversation with brings up the
   oft-mentioned "realish" (emphasis on the "ish") nature
   of his nonfiction memoirs, roll your eyes and say, "At
   least he's not James Frey." Then claim you think there
   should be a new genre for the type of memoir that
   Sedaris writes, something like the "might not be real
   but eat it up anyway because it's hilarious" memoir.

5. At the book signing you attended for *Me Talk Pretty One
   Day* Sedaris had a tip jar out on the table. Sedaris is the
   absolute best person to see while he's on book tour. If
   the person you're talking to doesn't already know that,
   inform them. Also mention how he once made smokers
   move to the front of the line because they have less
   time to live.

How to Fake Like You've Read J. D. Salinger

BASICS
You can't graduate from adolescence to adulthood without
reading J. D. Salinger's angst-ridden prose.

ESSENTIALOGRAPHY

Nine Stories

A collection of Salinger's nine best short stories, many of which originally appeared in *The New Yorker*.

Five words: Everyone's surprised by Seymour's death.

The Catcher in the Rye

The most (in)famous Salinger book. A youth upset at the world—for reasons undefined but generally because he is a youth—runs away to New York City.

Five words: Holden catches Phoebe, prostitute, phonies.

Franny and Zooey

How the Glass family copes with the death of the eldest brother, Seymour.

Five words: Glasses are tortured; attractive kids.

DETAILS

1. The most important thing to know about J. D. Salinger is that he continued to write up until his death, though none of this late work was published; supposedly he hoarded loads of stories, and they may still emerge in print. However, no evidence of these hidden manuscripts has come forward yet.

2. Some theorists say Salinger may have been publishing under pseudonyms. The person you're talking to might mention Thomas Pynchon as one of the theoretical

pseudonyms (this person is outdated in the conspiracy department and wrong).

3. As a Salinger fan, *The Catcher in the Rye* is not your favorite book. *The Catcher in the Rye* is the favorite book of people who haven't read anything else by Salinger.

4. If you're trying to impress a girl, say *Franny and Zooey* is your favorite book (it will make you seem sensitive). Say that you thought Lane was a douchebag.

5. If you're trying to impress a guy, say you love *Nine Stories* and then say, "'A Perfect Day for Bananafish' had such an insane and interesting end to it. It was the first great short story I ever read." If he heartily agrees with you, crack him up by proclaiming, "I see you're looking at my feet."

6. For being such a recluse, Salinger was a pretty big player. He famously went to court against ex-girlfriend Joyce Maynard to prevent her from publishing an autobiography about her relationship with him. He also was rumored to hook up with young girls who had written him fan mail.

How to Fake Like You've Read Jack Kerouac

BASICS

No person on the Phish tour would feel complete without a banged-up Kerouac paperback in their backpack. As one of the Beats, Kerouac had the coolest friends and they'd often appear in his works. Allen Ginsberg, Timothy Leary, and Ken Kesey were among those immortalized in Kerouac's books, which were often more truth than fiction.

ESSENTIALOGRAPHY

On the Road

Kerouac's thinly veiled semiautobiographical story of a young guy's adventures traveling across America.
Five words: Drugs, cars, trains, girls, drugs.

The Dharma Bums

A rambling, thinly veiled semiautobiographical account of Kerouac's quest to become a true Buddhist.
Five words: Outdoors, drugs, *yab-yum* sex, Buddhism.

Big Sur

Tale (yes, once again thinly veiled and semiautobiographical) of a popular author trying to sober up in a cabin.
Five words: A cabin, sex, drugs, drinks.

DETAILS

1. Pronounce Kerouac correctly. CARE (like the bear)
   o (like the face) ack (like that sound your mom made

when you accidentally sent those naked pictures of yourself to her instead of to your significant other).

2. Unlike Salinger fans, many true Kerouac fans can legitimately claim his most famous book, *On the Road*, as their favorite Kerouac book. *Big Sur* is often a close second. Scrutinize the person you are talking to. *On the Road* fans are a little less unkempt than those who are *Big Sur* fanatics. If someone's favorite Kerouac book is *The Dharma Bums*, step away slowly.

3. A surefire favorite quote from *On the Road* begins, "The only people for me are the mad ones . . ." Bonus points if you quote this next one as your favorite: "I like too many things and get all confused and hung-up running from one falling star to another till I drop." If the person you are talking to quotes that before you, watch out—you're wrestling with a real member of the literati.

4. Kerouac was bipolar and it was, as the myth goes, during a manic episode, reportedly high on Benzedrine, that he wrote all of *On the Road*. He wrote the entire manuscript on a thirty-one-foot-long scroll. Though it was whittled down extensively by Kerouac's editor—due to the graphic content of the book—there are editions available based, publishers claim, on the unredacted version of the scroll.

5. Talk about how you can't believe the amount of smoking and drinking that goes on, and if you want an easy out, segue into a conversation about *Mad Men* and how much that show makes you want to drink old-fashioneds. If you are unable to do so, move on to the points below.

6. Early in his twenties, Kerouac became close with many of the people who would become the figureheads of the Beat generation. So close, in fact, that he went to jail with William S. Burroughs when mutual friend Lucien Carr stabbed and murdered a man who was allegedly stalking him and confessed the crime to both men. Burroughs told Carr to turn himself in, while Kerouac took Carr to a movie and helped him dispose of the knife.

7. Say *On the Road* inspired you to take a road trip with friends (bonus points if you say you did it alone). *Big Sur* inspired you to go camping. *The Dharma Bums* inspired you to smoke more pot. *The Subterraneans* is why you started listening to jazz.

How to Fake Like You've Read Ayn Rand

BASICS

Very possibly Patrick Bateman's favorite novelist, Rand used her philosophy of objectivism to further her belief that an efficient society is one where people focus on benefiting themselves.

ESSENTIALOGRAPHY

We the Living

Semiautobiographical novel that first posited Rand's anti-communist views.

Five words: No power to the people.

The Fountainhead

Romantic, political novel about the importance of sticking to your ideals.

Five words: Hot sex scene with ginger

Atlas Shrugged

Atlas was Rand's magnum opus, her ultimate proclamation against collectivism.

Five words: Dagny sleeps with powerful men.

DETAILS

1. Let's start with the basics. Ayn Rand was a woman and a Soviet-born American immigrant. Her first name

is pronounced EYE-N. Easy to remember because it comes from the Hebrew word for eye.

2. You first read *Atlas Shrugged*, then *The Fountainhead*. Amazed by both, you went back and read *We the Living* and *Anthem*. Sigh and say, "But nothing compared to reading *Atlas Shrugged* for the first time. I take note whenever the date is September second."

3. Joke that you felt the last scene in *Atlas Shrugged* was so black and white that it could have been lifted from a comic book—all the freewheeling businessmen (and Dagny!) versus the evil government suppressors when breaking into the building containing Project F.

4. Say you used to play a drinking game based on *Anthem* when you were in college. The only rule is you can't say "I." If you do, you have to drink. Suggest that as a drinking game for the whole party.

5. Rand termed the philosophy underlying her books objectivism. Boiled down, Rand didn't give a fuck about others and didn't think you should either. Live for yourself. I know what you're thinking, and yes, Rand fans are great lovers. Just don't date them.

6. Fun fact? Rand had a six-foot-tall floral arrangement in the shape of a dollar sign next to her casket. Yeah.

How to Fake Like You've Read Arthur C. Clarke

BASICS

Don't write him off as someone only World of Warcraft fans pick up—Clarke was the first science fiction writer to receive a three-book publishing deal and shares an Oscar with Stanley Kubrick for the script of *2001: A Space Odyssey*. His work was both commercially and critically successful and he's credited with popularizing the idea of the telecommunications innovation geostationary orbit—he was even once nominated for a Nobel Peace Prize.

ESSENTIALOGRAPHY

2001: A Space Odyssey

Written alongside the production of Stanley Kubrick's film version, this doomsday tale of technology freaked out readers with its slow unraveling of a terrifying tale.

Five words: Hal's like a shady ex.

Childhood's End

A seemingly peaceful takeover of the world by aliens turns terrifying as they slowly brainwash children into becoming a collective consciousness controlled by their master, the Overmind.

Five words: Like *Independence Day*, only eerier.

Rendezvous with Rama

The first in a four-part series about the starship *Rama*, this book describes the ship's ominous appearance in the solar system and the mission dispatched to explore it.

Five words: Without characters or plot; horrifying.

DETAILS

1. Don't mix up Clarke with the other two science fiction authors, Isaac Asimov and Robert Heinlein, who round out the "Big Three." Heinlein and Clarke had a mostly cordial relationship throughout the years, but Clarke and Asimov had a famously playful yet cutting public relationship. Their glee at insulting each other led them to release a mock document called "The Asimov-Clarke Treaty of Park Avenue" stating that Clarke was a better science fiction writer and Asimov was a better science writer.

2. Talk about how surprised you were to learn that Clarke never won a Hugo (the science fiction writer's equivalent of an Oscar) for *2001: A Space Odyssey.* Then quickly correct yourself by positing that it was probably ineligible because Clarke wrote the book at the same time he was writing Stanley Kubrick's screenplay.

3. Clarke presents three "laws" in his works. Do not try to remember all three. Say this: "Because of Clarke, I've been able to predict the ending to every *Jurassic Park* movie and book." If pressed, say that the scene in *Ghostbusters* where they cross the streams reminds you of the first law even though Dan Aykroyd wasn't

that old. (The first law is: When a distinguished but elderly scientist states that something is possible, he is almost certainly right. When he states that something is impossible, he is very probably wrong.)

How to Fake Like You've Read Don DeLillo

BASICS

Don DeLillo has influenced more than his fair share of budding authors. His most revered work is *White Noise*, a slim volume compared to his magnum opus *Underworld*, but still a behemoth fixture in the postmodernist canon.

ESSENTIALOGRAPHY

White Noise

Unlovable characters abound in this slow creep toward destruction.

Five words: Hitler professor realizes contemporary doom.

Mao II

An American in fear of terrorism before Americans knew terrorism.

Five words: Quote, "Future belongs to crowds."

Underworld

A half century of American social evolution and decay fills this grand, meditative tome—the prologue about a baseball game between the Brooklyn Dodgers and New York Giants is often cited by reasonably athletic yet literary young men as their favorite piece of American fiction.

Five words: No plot. Eight hundred pages.

Point Omega

More of a social tract than a novel, DeLillo makes his anti-war stance known through his main character, a young American filmmaker in Iraq.

Five words: *24 Hour Psycho*; humming anxiety.

DETAILS

1. You can't bring up DeLillo without mentioning the famous story of the business card emblazoned with the phrase "I don't want to talk about it," which he would hand out to reporters when prompted to do just that. Eventually, a reporter caught up to him in Greece. DeLillo agreed to an interview and has acquiesced to more since but remains relatively antisocial. In one interview he responded to questions by asking, "Is it dinnertime yet?"

2. "The future belongs to crowds" (from *Mao II*): A speaker at your college graduation actually invoked this line in an inspired attempt to rouse the graduating class to set out into the world and promulgate the positive beliefs of the university, with the help of the alumni association. The speaker obviously had no idea what he was referencing, as DeLillo's theme revolves

around the idea that the collective identity diminishes the power of the individual. The only thing keeping you in your seat that day was the knowledge that your grandma had flown in to be there.

3. It is said that DeLillo used to keep two files on his writing table, labeled "Art" and "Terror." In a self-deprecating way say that your two files would be "Art" and "Tweets."

4. Mention how much you loved the *McSweeney's* essay about DeLillo as a stadium vendor. Mimic the cry of a peanut thrower while you say, "Hot dogs cloaked in foil wrappers!" Then mention how great *McSweeney's* is and how brilliant Eggers is for starting it—too bad his novels aren't any good.

5. Describe a site you once stumbled on that combined randomly generated fragments of *White Noise* into one coherent string. Express surprise at how lucid these outputs were and mention it's because of DeLillo's consistent voice that his works can be chopped up and rearranged and still make sense.

6. In 1998, DeLillo shilled for Oldsmobile. He was the face on their print ad in the *New York Times*. Other authors on the ad wagon: Hemingway posed for Ringling Bros. and Ballantine Ale beer ads, airlines, and

a pen company; John Steinbeck also lent his visage to Ballantine Ale.

How to Fake Like You've Read Jonathan Franzen

BASICS
Jonathan Franzen's legacy is not his influence on but rather his *intimidation of* the average writer. His lengthy tomes with social bents gleefully and mercilessly knock the shit out of any short story collection.

ESSENTIALOGRAPHY
The Corrections
A mom's quest to have her unhappy and disillusioned family home for the holidays.
Five words: Detailed Americana incarnate (fuck Oprah).

How to Be Alone
Franzen's collection of essays, the most famous of which decries the death of the social novel.
Five words: Fuck the contemporaries; "Why Bother?"

Freedom
A loveless and broken family deals with middle-class problems.
Five words: Forget fucking Oprah, publicity whore.

DETAILS

1. The oft-mentioned and unavoidable Franzen anecdote is his public dissing of Oprah. In an interview shortly after being selected for Oprah's Book Club, he expressed concern that the "Oprah's Book Club" sticker on the cover of his books would repel male readers. Oprah responded by pulling the already-produced interview she had done with him.

2. Franzen releasing a book is the equivalent of one hundred thousand simultaneous media journalist and blogger orgasms. If you are not a fan of Franzen, your best bet is to stay off the Internet entirely for the month before and after the release of his work, in order to be spared everything from over-the-top Franzen praise to admonishments from female authors. Most famously, popular chick lit authors Jennifer Weiner and Jodi Picoult chastised the media for writing off their novels as fluff and exalting Franzen as a god among men (the *New York Times* reviewed Freedom twice in one week).

3. Make a remark about how closely your family drama mirrors Jonathan Franzen's: sometimes painfully dull, sometimes flashing with drama. Franzen's real gift is creating characters who are relatable to all. It doesn't matter what your family is actually like; it mirrors Franzen's families.

4. Franzen was so serious about avoiding distraction while typing up *The Corrections* that he blindfolded himself, put in earplugs, and donned earmuffs in his soundproof writing studio.

5. As newlyweds, Franzen and his now ex-wife, writer Valerie Cornell, used to write for eight hours a day and after dinner read for five more. Their wedding anniversary was the only day they'd go out to eat.

6. Franzen refuses to explain the meaning behind the title of *Freedom*. *The Corrections* was so named because originally it dealt with prison programs.

How to Fake Like You've Read Ian McEwan

BASICS

This English novelist has been quoted as saying his goal in writing is to "incite a naked hunger in readers," but mainly his writing makes me frightened to get naked. Sadomasochistic themes abound in his writing and keep the reader adequately piqued.

ESSENTIALOGRAPHY
Enduring Love
The haunting story of a man who begins to stalk a couple after their lives intersect in a fatal accident.
Five words: Think more frightening *Fatal Attraction*.

The Comfort of Strangers

A disturbing tale that will make you think twice about mingling with locals in foreign countries as the protagonist couple are swept into a horrible predicament by a seemingly endearing stranger they meet while touring Venice.

Five words: Sadomasochism, cameras, and drugged tea.

Atonement

A young woman falls in love with a man only to be thwarted by her confused sister's claim that he is a rapist.

Five words: Postmodern twist ending explains title.

DETAILS

1. Jonathan Franzen and Ian McEwan had a years-long feud that appears in interviews to be more media hype than actual strife. Most point to McEwan's comment after John Updike's death that Philip Roth was the last great living American novelist. Franzen countered by including a bit in *Freedom* about a character's not being able to get very interested in *Atonement*. However, in interviews after the release of *Freedom* Franzen spoke well of McEwan, noting that he sent him a galley of the book early on and that McEwan was a "really nice guy" about it.

2. His mother's name was Rose Lilian Violet. First wife was named Penny, second wife was named Annalena. Most of his female characters also have excessively flowery names: Briony, Cecilia, Lola, Rosalind, Daisy.

3. A rumor circulated that he only wrote fifteen words a day, thanks to a journalist mistakenly transcribing "fifteen words" instead of McEwan's correct answer: five hundred words per day.

4. Following a bitter divorce, a court granted McEwan full custody of his children. His ex-wife then kidnapped their thirteen-year-old son and took him to France. The child was returned within the day. In another bit of familial turbulence, McEwan found out as an adult that he had a brother: his parents had had an affair when his mother was still married to her first husband, and they gave the child up for adoption. When her husband died in World War II, McEwan's mother was free to marry the man she'd had the child with, David McEwan.

How to Fake Like You've Read Norman Mailer

BASICS

My mother claims she saw Norman Mailer in person once when she was in the Atlanta airport with me and my siblings while trying to make a connecting flight to Texas. She said he was wearing a crumpled linen suit and carrying a beat-up briefcase. He was probably on his way back from a movie set or a stump speech or a book tour: Norman Mailer attempted to be everything from a mayoral candidate to a movie director, and failed at most.

ESSENTIALOGRAPHY

The Naked and the Dead

Mailer's first novel deals with soldiers in the South Pacific during World War II. It famously substituted "fug" for "fuck" at the request of the publisher.

Five words: Fug clemency and fug compassion.

Ancient Evenings

Heavily panned by critics and the public, Mailer spent about a decade writing this mammoth, which manages to be more sex than plot or character development.

Five words: Foot fetishists will love it.

Harlot's Ghost

A book with more words than David Foster Wallace's *Infinite Jest*, this novel about the life of a CIA agent mixes genres and aggrandizes characters in a transparent and futile attempt to create an epic novel.

Five words: "To be continued," please don't.

DETAILS

1. Want to see something bone-chilling? Search online for the video of Rip Torn attacking Normal Mailer with a hammer on the set of the movie *Maidstone*. Around the same time, Mailer also stabbed his wife with a penknife after she called him "faggot" at a party. His other notable fisticuffs include punching Gore Vidal in the greenroom of *The Dick Cavett Show*. The legend

goes that Vidal, still lying on the floor, said, "Words fail Norman Mailer yet again." In another feud, Truman Capote called Mailer talentless. In response, Mailer sat on him.

2. At his death, the *New York Times* ran an obituary titled "Norman Mailer, Towering Writer with Matching Ego, Dies at 84." The obituary makes mention of his several overheated attempts to create the "Great American Novel," which would likely be the most embarrassing attribute for any author to have committed to posterity after death. Still, Mailer's ego could probably take such a blow, as the obituary reveals the source of his confidence: when he was a child, his mother used to repeatedly, obsessively tell him he was "perfect."

3. Mailer was a cofounder of the *Village Voice* but didn't make much of an impact at the paper. Reports say his articles were often turned in late and in need of intense editing.

4. Mailer believed in reincarnation. He wanted to come back as a black athlete but he said that with his luck, he'd return to earth as a cockroach. Keep that in mind next time you find one in your bathtub.

How to Fake Like You've Read Charles Bukowski

BASICS

It's not often I can say plainly that a novelist and poet's dominant themes are sex and booze and being dirty, but with Bukowski, that's where it's at. Typically read at a slightly later phase and age than J. D. Salinger. Whatever you do, don't listen to anyone who connects Bukowski to the Beats. He despised them and rather was massively influenced by Italian-American Los Angeles–based author John Fante, whom he has cited as his "God."

ESSENTIALOGRAPHY

Post Office

A friend offered Bukowski, who was employed at the time by the post office, $100 per month for life if he'd quit—he did and *Post Office* was written shortly after. Mostly about drinking, gambling, and lusting after women while working at lowly jobs.

Five words: Semiautobiographical dirtiness in jaded adulthood.

Notes of a Dirty Old Man

Bukowski's articles detailing his semiautobiographical, alcohol-drenched adventures.

Five words: Drinking, sex, dirty apathetic fame.

Pulp

Dramedy or parody (depends on who you ask) of detective novels, as done by a grimy old man.

Five words: "Dedicated to bad writing": accomplished.

DETAILS

1. "Don't try," a line from one of his poems, is written on his gravestone. Bukowski believed writing should come simply, from the gut.

2. Bukowski wrote the first poetry you could stomach. Forget T. S. Eliot; nothing piqued your tenth-grade sensibilities more than "Girl in a Miniskirt Reading the Bible Outside My Window" or "My First Affair with That Older Woman."

3. Bukowski wrote what he knew. The cat in *Ham and Rye* is his cat. The main protagonist of several of his works, Henry Chinaski, is undoubtedly Bukowski himself. From their first names (Charles's real first name is Henry) to their love of the racetrack, their job history, and their women, the only big difference between the two was Bukowski's habit of moderating himself every once in a while, trying out celibacy for years, or cutting back on drinking periodically.

4. Main protagonist Chinaski can be compared to that one particularly dirty ex-boyfriend or roommate whom you took a chance on and let move in with you, thinking the arrangement wouldn't be too bad. Waxing poetically about shit stains in his underwear, advising us that "sometimes you have to pee in the sink," Bukowski made no excuses for Chinaski's skid row lifestyle, which makes you love him more. An observation from *Ham on Rye*: "There was nothing really as glorious as a good beer shit—I mean after drinking twenty or twenty-five beers the night before. The odor of a beer shit like that spread all around and stayed for a good hour-and-a-half. It made you realize that you were really alive." I leave you with that.

How to Fake Like You've Read Deborah Eisenberg

BASICS

This austere short story writer (and onetime playwright) is a critics' darling, though she is not as well-known as her contemporaries. We have big tobacco companies to thank for her entrance into literature; Eisenberg started writing as a way to divert her attention while she quit smoking. This is confusing to most of us because a cigarette is usually exactly what we need before we can set our fingers to keys.

ESSENTIALOGRAPHY

Transactions in a Foreign Currency

A short-story collection trailing young women through their dissatisfying relationships.

Five words: Reasons to avoid Canadian men.

Under the 82nd Airborne

A story collection expanding from settings in Central America to New York City with mostly female protagonists exploring harsh truths about their lives.

Five words: Forget narrative, focused on details.

Twilight of the Superheroes

Varied stories all rotating around the lives of young people in post-9/11 New York City.

Five words: Dispirited, nuanced insanity and introspection.

DETAILS

1. Eisenberg has been together with Wallace Shawn for over thirty years. Does that name ring a bell? He played Vizzini in *The Princess Bride* and Mr. Hall in *Clueless*. More substantially, he's an author and playwright of very well-received works. He's also the son of William Shawn, longtime and revered editor of *The New Yorker*, to whom J. D. Salinger's *Franny and Zooey* is dedicated.

2. In interviews Eisenberg seems to present herself as a lazy genius, with a self-deprecating style that stresses

her lack of ambition. You can say her interview style is purposefully off-putting and that you haven't seen anyone try so hard to make it seem like they don't care.

3. The first story in *Twilight of the Superheroes* almost too succinctly captures the relationship you had with your first set of friends postcollege.

How to Fake Like You've Read John Updike

BASICS

Hailing from a small town, this tall writer was the spitting image of his main protagonist, Harry "Rabbit" Angstrom. The prolific author of more than thirty works characterized his writing as giving "the mundane its beautiful due" and, you'll agree, his writing is stunningly lyrical at moments when he's describing scenes that the rest of us take for granted as part of everyday life. However, I don't include his sex scenes in that assessment (more on that farther down).

ESSENTIALOGRAPHY

Rabbit, Run; Rabbit Redux; and *Rabbit at Rest* (three out of seven in the Rabbit series)

The Rabbit series is the examination of a man named Rabbit Angstrom stuck in his way of life, the victim of social constraints and a small town that offers very little in the way of meaningful stimulation. Most interesting point in *Rabbit, Run*: Rabbit's newborn daughter is accidentally drowned in the bathtub by his manic-depressive and often drunk wife. Most interesting point in *Rabbit Redux*: A teenage girl becomes his lover but eventually dies when Rabbit's house burns down. Most interesting points in *Rabbit at Rest*: Rabbit has a one-night stand with his son's wife; his longtime mistress dies; he dies.

Five words for *Rabbit, Run*: Rabbit as bored young father.

Five words for *Rabbit Redux*: Rabbit as a middle-aged man.

Five words for *Rabbit at Rest*: Rabbit as fat old man.

Couples

Adultery and deceit abound in the ordinary lives of people living in a quaint town during the 1960s.

Five words: To marry is to dissatisfy.

The Witches of Eastwick

A group of women take on magic after their marriages fail and are charmed by a dastardly gentleman.

Five words: Movie version gave me nightmares.

DETAILS

1. Updike's examinations of small-town monotony are fairly tame until he gets to the sex scenes, which are awful. *The Widows of Eastwick* garnered him bad-sex-in-fiction prizes thanks to lines like "Her face gleamed with

his jism in the spotty light of the motel room, there on the far end of East Beach, within sound of the sea."

2. He was such a prolific writer because he swore by his routine of writing at least three pages every day. Bring that up any time an acquaintance bemoans how hard it's been for them to finish their book on deadline.

3. Updike acknowledged Bret Easton Ellis's *Less Than Zero* in the afterword to *In the Beauty of the Lilies*, then in subsequent interviews admitted Ellis's novel wasn't that great—its main message being simply "to get wasted and stay wasted"—but that it had helped him formulate his own writing on "the sort of burnt-out generation of Hollywood–Los Angeles kids." You can cite this as one of the reasons Bret Easton Ellis has such a chip on his shoulder.

4. *The Centaur*, Updike's second National Book Award–winning novel, isn't as well-known as the Rabbit series but is a must-read among his die-hard devotees. If you want to offend a major fan, say, "Updike's egregious use of metaphor masks his chronically underdeveloped plot." If you want to impress them say, "Metaphors allowed him to dwell within the abstract beauty of everyday happenstance." Same meanings, different connotations.

How to Fake Like You've Read Bret Easton Ellis

BASICS

Ellis's writings define contemporary novels set in Los Angeles. No author can avoid the comparison if they choose Hollywood as their backdrop. His characters' California-It-crowd eighties mind-set awoke millions of teenage yearnings for the West Coast, regardless of his characters' frequently dismal fates.

Ellis was influenced by and influenced his Bennington College friends and fellow writers Jonathan Lethem, Jill Eisenstadt, and Donna Tartt. Much ink has been spilled debating Ellis's possible allusions to his peers in *The Rules of Attraction* (his novel about a group of college friends). Critics in the 1980s also repeatedly lumped Ellis into literature's version of the "Brat Pack," which included Tama Janowitz and Jay McInerney.

ESSENTIALOGRAPHY

Less Than Zero

Drugs and prostitution drive the story of a group of young Beverly Hills scenesters.

Five words: Too wealthy twentysomethings' ennui experiment.

American Psycho

Perennial favorite of every well-dressed, slightly psychotic male who thinks autoerotic asphyxiation is actually kinda hot, this transgressive novel describes the life of a character who is Mr. Perfect Businessman by day and a serial killer by night.

Five words: Don't trust too pretty men.

Imperial Bedrooms

Follow-up to *Less Than Zero* with older incarnations of many of its characters; Ellis gives readers a glimpse into an L.A. casting call.

Five words: American Apparel revelations; falls flat.

DETAILS

1. Ellis publicly proclaimed exactly what was on your mind when J. D. Salinger died. He tweeted, "Yeah!! Thank God he's finally dead. I've been waiting for this day for-fucking-ever. Party tonight!!" You were never a fan of Ellis until this moment. Seemingly inappropriate to the uninitiated, it was in fact exactly the kind of sentiment Salinger would have expressed himself.

2. *American Psycho* was the only book your fratty friend bothered to bring to college; he'd get drunk and read aloud the particularly gross parts. Completely missing the critique of superficiality, materialism, and hedonism, he strove to someday own suits and watches as glamorous as Patrick Bateman's.

3. Ellis modeled Patrick Bateman after his abusive father and has publicly remarked that his desire to succeed probably came from the need to "prove something to Daddy."

4. Ellis also has observed that younger women are much bigger fans of his work than older women: once a young woman in her twenties had him sign two copies of *American Psycho* and then whispered to him, "This book taught me how to masturbate." You can claim the same, naturally.

5. Ellis's fictional world is fluid: characters from one book pop up in another. Sometimes even characters from his friends' books make appearances. Most notably: Jay McInerney's drug-and-sex-addicted protagonist Alison Poole from the novel (and rumored roman à clef) *Story of My Life* is sexually assaulted by Patrick Bateman in *American Psycho* and is a girlfriend to the main protagonist in *Glamorama*. Alison Poole is based on Rielle Hunter, lover and baby mama of John Edwards; she once dated McInerney.

6. Ellis delivered this opinion of David Foster Wallace after his erstwhile competitor and colleague's suicide: "The journalism is pedestrian, the stories scattered and full of that Mid-Western faux-sentimentality and *Infinite Jest* is unreadable." Blame his bitterness on sore feelings in light of Updike's public statement that Ellis's books were noteworthy only as evidence of a generation's moral bankruptcy.

How to Fake Like You've Read Sarah Vowell

BASICS

A bubbly, nasally voice is the most defining characteristic of this nonfiction author, who is probably talking on NPR as you read this.

ESSENTIALOGRAPHY

The Wordy Shipmates

The adventures of the first settlers in the Massachusetts Bay Colony.

Five words: *New York Times*: epic pan.

Assassination Vacation

Offbeat examination of the execution of American presidents told through a road trip (or as she reverentially called it, a "pilgrimage") to all the assassination spots.

Five words: *New York Times*: lukewarm review.

Unfamiliar Fishes

Freewheeling history lesson about Hawaii's tourist highlights.

Five words: *New York Times*: beyond condescending.

DETAILS

1. To understand Sarah Vowell is to have heard her voice. She was the voice of Violet in the Disney movie *The Incredibles*. She's consistently described as an "NPR

darling." Her boss is Ira Glass of *This American Life*. Fans call her hyperarticulate. Critics call her grating. One of her biggest critics, Virginia Heffernan at the *New York Times*, observed that she writes like she talks. It might be refined in oral performance but it doesn't translate well into print.

2. Her nonfiction books are preoccupied with historical trivia, which readers are bound to find either trite or engrossing. Fans of Vowell are likely faking their seemingly expansive historical knowledge as much as you're faking your literary mastery with the help of this book. Ask them about their interest in the French Revolution to see their faces fall flat. Similarly, you're screwed if they want to initiate a dialogue on differences between Richard Yates's *Revolutionary Road* and *Young Hearts Crying*.

3. You most admire Vowell for her open admission that she cuts her own hair to avoid having to talk to a hairstylist. Being elitist is so cumbersome. Also, she doesn't drink coffee, which is weird. If you told me that she didn't drink liquor as well, I would have to insist that she must not write her own books because I have no idea how to write without one or the other.

How to Fake Like You've Read Alice Munro

BASICS

Short-story writer who presumably took her own experiences as an unhappy Canadian homemaker as fodder for her fiction. At the center of her stories, almost invariably, is a Canadian housewife, consumed with her household but unthanked by her family and stymied in her aspirations.

Influenced by: Every review of Munro mentions that she is our generation's Chekhov but she often brings up Eudora Welty as an influence.

ESSENTIALOGRAPHY

Dance of the Happy Shades

Munro's first published collection of short stories, in which she introduced her oft-recurring discontented small-town female character.

Five words: "Red Dress," required pre-prom reading.

The View from Castle Rock

A fictionalized account of true events from Munro's personal life and ancestral family history.

Five words: Storified memoir; family tree ponderings.

Too Much Happiness

Short stories that present a macabre twist on the classic Munro housewife protagonist.

Five words: "Men carry nothing; women all."

DETAILS

1. If you're ever caught having only partially read one of her books, take comfort in Munro's belief that one should inhabit the world the book presents and wander at will rather than read the entire thing from start to finish.

2. Munro, with her first husband, used to own a bookstore and work the cash register. Her now-ex-husband still owns the bookstore in British Columbia, called Munro's Bookstore. Before being published, she would hide her attempts at writing as if they were shameful journal entries, for fear she'd be mocked by readers.

3. She refers to a short story as a "chunk of fiction." Her abrupt endings make you think of them more as chunks with jagged edges. Once in Paris I was walking down a street and a large, dense object fell from God knows where and onto my head. I lifted my hand to brush it off and looked at the writhing creature I had struck onto the ground. It was a (small) rat, struggling to get back onto its feet. The intensity with which I freaked out, rubbing my hair all around my head to make sure I had gotten all of the possible debris the rat carried with it out while screaming, "A rat! A rat!" in English to passing Parisians who seemed to not understand what had just happened to the crazy American girl—that's how Munro endings can feel.

Like an inexplicable moment you just got pulled
through, that no one nearby saw happening.

4. If you have ever wondered how Munro so fully grasps
   and succinctly captures the plight of the small-town
   homemaker, you need look no farther than her own
   path as a writer for an explanation. HOUSEWIFE FINDS
   TIME TO WRITE STORIES, read the headline in her local
   newspaper when it was discovered that she had been
   published.

How to Fake Like You've Read William Faulkner

BASICS

The touchstone of Southern Gothic fiction, this novelist and
onetime poet changed contemporary American literature
for the better.

Influence by: Faulkner's biggest influence was his great-
grandfather Colonel William Clark Falkner (the "U" was
added in subsequent generations), whose life supplied the
template for several of Faulkner's characters (and who was
also a writer himself). It is said that as a child Faulkner once
proclaimed, "I want to write like my great-grandaddy."

ESSENTIALOGRAPHY

The Sound and the Fury

Connected stories from four different points of view, all detailing the downfall of Caddy Compson.

Five words: Dilsey saved all their asses.

Light in August

Another pregnant girl, another Yoknapatawpha County backdrop.

Five words: Overstated symbolism of Joe Christmas.

As I Lay Dying

Dead mother's sons carry her coffin to her final resting place.

Five words: "My mother is a fish."

DETAILS

1. You can joke about the time in college a fellow student compared *As I Lay Dying* to the film *Weekend at Bernie's*. Faulkner's novel revolves around the harrowing journey a family must take to deliver the casket bearing their deceased mother, Addie, to its final resting place. The student used the repeated references to the putrid smell of Addie's rapidly decaying body to expose a major oversight in the campy antics of *Weekend at Bernie's*.

2. Whenever someone compliments you for correctly connecting the dots in a complicated situation,

sarcastically mention that you're a big Faulkner fan.
Faulkner readers are often left with the feeling that
they've been suddenly and abruptly thrown into the lives
of ten different characters. People who don't like Faulkner
are typically too ADD to stay with the arduous narratives,
shifting points of view, and long sentences. Faulkner's
advice to those who can't understand his points even after
rereading two or three times was to "read it four times."

3. Most of his novels take place in a fictional Southern district
   named Yoknapatawpha County. Faulkner drew detailed
   maps of how it was laid out in his mind. Yoknapatawpha
   is pronounced pretty phonetically: yahk-na-paw-TAW-fa.
   The word comes from a portmanteau Faulkner created
   of the Chickasaw words "split land." The first work in
   which Yoknapatawpha County appeared was *Sartoris*,
   which revolved around Colonel John Sartoris, a character
   modeled after the great-grandfather mentioned above.
   Colonel Falkner had built the railroad that Faulkner's
   father worked on as a conductor throughout his life.
   Faulkner one-upped them both by creating a whole town.

4. Ernest Hemingway and William Faulkner had an
   ongoing feud that was rooted, fundamentally, in their
   different writing styles. Faulkner accused Hemingway
   of never using a word that would make a man open
   a dictionary and Hemingway retorted that big
   words aren't necessary for big emotions. Their work

converged when Faulkner contributed to the screenplay of the film adaptation of Hemingway's *To Have and Have Not*. The famous line "You know how to whistle, don't you, Steve? You just put your lips together—and blow," is all Faulkner.

How to Fake Like You've Read Toni Morrison

BASICS

The figurehead of contemporary African-American literature, Toni Morrison is the go-to favorite for anyone who gets their political enlightenment through Oprah.

ESSENTIALOGRAPHY

Sula

A town is united through its hatred for one inhabitant, Sula, but after her death the townspeople revert to discordant lives. Most of the story revolves around Sula's friendship with and estrangement from another woman and the inter-racial relationships that she carries on with no regard for social mores.

Five words: Lesson isn't hate harmonizes, right?

The Bluest Eye

Heart-wrenching tale of incest, alcoholism, rape, death, and every other thing that makes life shit.

Five words: "Wicked people love wickedly." —Claudia

Beloved

Mother and daughter escape slavery only to find themselves haunted by the ghost of the mother's murdered daughter.
Five words: Repression manifests in perceived freedom.

DETAILS

1. Toni Morrison famously called Bill Clinton "the first black president"—qualifying his background (single-parent household, poor, Southern) as "black." As a result, she was limited to labeling Barack Obama "the man for the time" when she endorsed his candidacy for president over Hillary Clinton.

2. Morrison's birth name was Chloe Ardelia Wofford, which she changed to Chloe Anthony Wofford at age twelve when she converted to Catholicism. While at Howard University, friends began calling her "Toni," the name of a popular hair permanent at the time.

3. Morrison had her fair share of ill fortune in the wake of her Nobel Prize triumph: her house burned down shortly after the award presentation but not soon enough that she hadn't already deposited the prize money in a retirement fund she couldn't touch. Morrison was the first black woman to receive the Nobel Prize in Literature. On her feelings about winning the Nobel: "I felt blacker than ever. I felt more woman than ever."

4. Morrison purposely avoids plots involving much confrontation between whites and blacks because she feels that too much of black literature already deals with conflict between the races. To her, the black community is insular and worthy of her novels' primary focus.

5. You can bring up *Sula* whenever some girl says, "It's much easier for me to be friends with guys than girls." *Sula* was Morrison's 1971 attempt to illuminate a deficiency of female protagonists in contemporary literature. In *Sula*, Morrison consciously created a novel with two female friends at its center. Decades after the release of *Sula*, Morrison noted that the decision to focus on female relationships was becoming a conventional one, "and it's going to get boring. It will be overdone and as usual it will all run amok." You can say *Sex and the City* is the best example of this concentration on female relationships run amok, per Morrison's prediction.

How to Fake Like You've Read Haruki Murakami

BASICS

Japanese novelist who redefined Japanese literature by blending it with Western themes, among them absurdism and existentialism. His love of jazz music and popular culture punctuates his writing, as characters often listen to carefully selected tracks and several works are titled after songs. Murakami used to run a jazz club and owns over seven thousand vinyl records.

ESSENTIALOGRAPHY

Kafka on the Shore

This "riddle as story" follows the intersecting lives of two characters: Kafka, a young boy who has run away from home, and Nakata, a mentally impaired man who can talk to animals.
Five words: Oedipal themes and raining fish.

Norwegian Wood

Western-pop-culture-addled recollections of a thirty-seven-year-old Japanese man's time in college and various dramatic love triangles.
Five words: Crazy girl always gets attention.

The Wind-Up Bird Chronicle

Self-actualization occurs for a slacker when his quiet life is disrupted by his runaway cat.
Five words: Existentialist plot; goal is self-discovery.

DETAILS

1. Don't confuse Haruki Murakami with Takashi Murakami—both are prolific and beloved by the intelligentsia. Takashi, however, is an artist—most celebrated for establishing "Superflat" as a painting technique. Superflat is the melding of anime with

incisive messages on the superficiality of modern culture.

2. Murakami claims he never read Japanese literature while in school: "If I'd read Japanese literature," he says, "I would have had to talk about it with my father, and I didn't want that." You can adopt this excuse for why you never picked up William Faulkner or Ernest Hemingway.

3. You could describe the inverted dynamic of *Kafka on the Shore* and *The Wind-Up Bird Chronicle* as unrelatable characters with a relatable sentiment in the former versus relatable characters with an unrelatable sentiment in the latter.

4. Murakami attributes his decision not to have children to the fact that he lacks the sort of confidence his "parents' generation had after the war that the world would continue to improve." Repeat that at a holiday dinner when your mother is needling you on when you're going to give her some grandkids.

5. He's an avid runner—such an avid runner in fact that he titled his memoir *What I Talk About When I Talk About Running*, and once said that he wanted his epitaph to read "At least he never walked." If you're an indolent type you can say you aspire to Bukowski's "Don't try" inscription.

How to Fake Like You've Read Philip Roth

BASICS

The modern characterization of the Jewish American man springs from much of this novelist's work.

Philip Roth was inspired by his own life. He is best known for protagonists who are thinly veiled stand-ins for the author himself. (His first wife also appears in several of his stories, including *My Life as a Man*, in which a scheming girlfriend tricks the protagonist into marrying her by purchasing urine from an impoverished pregnant woman to fake a positive pregnancy test.)

ESSENTIALOGRAPHY

Goodbye, Columbus

His National Book Award–winning first work, a collection of stories about young Jewish Americans struggling to find their identity in the 1950s.

Five words: Librarian meets Jewish American Princess.

Portnoy's Complaint

This tale of adolescence put Roth on the map as an obscene and comedic writer, thanks particularly to a masturbation scene involving a raw liver (which puts anything from *American Pie* to shame).

Five words: "Fucked my own family's dinner."

American Pastoral

Pulitzer Prize–winning reflection on 1960s America through the eyes of Roth's favorite protagonist, Nathan Zuckerman. **Five words:** The Swede's Merry isn't merry.

DETAILS

1. "Wee-quake" is how to pronounce the neighborhood Roth hails from—the spelling is Weequahic. This working-class area's influence plays an important role in much of his work.

2. The two authors whom you would love to get a drink with or have over for dinner have to be Saul Bellow and Philip Roth. These two, who were close friends, carried on a lengthy correspondence, much of which was mutually congratulatory and wonderfully inappropriate, as in this excerpt from Bellow to Roth: "If I had been interviewed by an angel from the Seraphim and Cherubim Weekly I'd have said, as I actually did say to that crooked little slut [from *People* magazine], that you're one of our best and most interesting writers."

3. For a while in summer 2008 you had the Philip Roth ringtone—it was a viral sensation after an interview where Roth pretended to mock "Jewish shouting" as exemplary of what the movie version of *Portnoy's Complaint* was like.

4. Roth has said, "When the whole world doesn't believe in God, it'll be a great place." Repeat this next time you see a story on the news about religious uprisings or an upcoming Tea Party event.

5. Next time the media is abuzz with winners of the Nobel Prize in Literature, moan, with profound exasperation, "Roth didn't win *again*?!" Roth's bald lust for the Nobel is widely known—more so even than Mailer's ambition to write the Great American Novel. Friends have attested to Roth's depression in the wake of tuning in to hear the winners announced. In more recent interviews, Roth says he tries not to even be aware of the day they announce the year's winners.

How to Fake Like You've Read Lionel Shriver

BASICS

Expat novelist Lionel Shriver, who now lives in London . . . ugh, God, I'm bored just writing this description. I think she's more interesting than she comes across in interviews, which are horribly safe and seem to be controlled by her to avoid exposing anything unique besides the standard sentiments that writing is hard work and sometimes it's a painful process. Her characters, however, are so vibrantly emotional, like tuning forks set to a high pitch, usually pushed beneath a calm countenance for the sake of keeping up appearances, that it makes me feel as if she's constantly writing herself into the narratives.

ESSENTIALOGRAPHY

We Need to Talk About Kevin

An epistolary novel exploring an egregious act of violence committed by a child who annoyed his mother and escaped the interest of his father.

Five words: Perfectly captured anxiety of Columbine.

The Post-Birthday World

Split into two parts, the book portrays how the protagonist's life would be if she had an affair and how it would proceed if she didn't.

Five words: "What if" world, *Sliding Doors*.

So Much for That

Gripping account of the damage terminal illness can wreak on a marriage and family.

Five words: Health-care bills become tragic fate.

DETAILS

1. Shriver is often photographed wearing heavy fleece gloves. Before you scoff at her habit as eccentric, she suffers from Raynaud's phenomenon, a disorder that causes fingers to hurt and turn blue or white from lack of proper circulation.

2. She legally changed her birth name, Margaret, when she was fifteen years old. Shriver has admitted to lying to reporters in interviews and claiming that her parents always wanted a boy. In reality, she thought it was a strong name and felt she deserved to go by something more powerful and therefore more masculine.

3. Cheer up any recently discouraged aspiring writers by letting them know Shriver's powerful *We Need to Talk About Kevin* was rejected over thirty times. Also, you can advise the same writer friends to avoid the debt-laden route of graduate school by quoting Shriver's scoff about her MFA degree. She declared that the accolade has "a kind of indulgent, middle-class gestalt."

4. Many interviewers broach the subject of Shriver's childlessness. They bring up the fact that the mother from *We Need to Talk About Kevin* has been described as the "Anti-Mom" and segue into questions about why the author hasn't had children. She cites her selfishness as the main reason. Focusing even more on Shriver's love and familial life, other interviewers home in on her marriage to her ex-agent's husband. Reading one of these borderline-intrusive interviews with Shriver makes it easy to understand why some writers are resolutely reclusive and publicity averse.

How to Fake Like You've Read Margaret Atwood

BASICS

Margaret Atwood is a Canadian novelist and poet often pigeonholed as a feminist writer, though her writings range from intense examinations of Canadian identity to highly political tracts.

ESSENTIALOGRAPHY

The Handmaid's Tale

In the future, a totalitarian government oppresses all of its citizens through visual classifications of their roles for the purpose of creating a more efficient society.

Five words: Men rule, girls . . . make babies.

The Blind Assassin

Half mystery, half love story, part novel within a novel—the story of a man on the run, the woman he loves, and the sister in the middle.

Five words: Rote descriptions with grandiose plot.

The Year of the Flood

Loosely connected to Atwood's previous work *Oryx and Crake*, a nature cult called God's Gardeners predicts a fatal plague that wipes out much of Earth's inhabitants.

Five words: Stripper, waitress, liobams for salvation.

DETAILS

1. Despite Margaret Atwood's horrid Twitter feed, you can't bring yourself to unfollow her. As a fan, to know that she's expounding thoughts into the atmosphere without seeing them is inconceivable, no matter how many boring retweets and links she posts.

2. If Canadian politics is a helpful reference point for you, Atwood is a Red Tory, not a Liberal. She's very explicit about this in interviews. Mainly because she believes that power should reside in the hands of the community rather than the moneyed establishment. Her principles are reflected in the dystopian framework of her novels.

3. Atwood describes *Oryx and Crake* as "adventure romance"; use that line next time someone speaks condescendingly about science fiction.

4. If you're a woman, you first heard about masturbation while reading *The Handmaid's Tale.* And no, that's not why "handmaid" is in the title.

How to Fake Like You've Read Tao Lin

BASICS

This egregiously twee novelist and poet whose biggest accomplishment is his social media presence also runs Thought Catalog, a blog with a mission to let amateur writ-

ers emote about twentysomething middle-class problems.
Lin has been relegated to the genre "urban hipster lit."

ESSENTIALOGRAPHY
Eeeee Eee Eeee
Eeeee Eee Eeee affirmed his agenda as annoyer provocateur.
Five words: This book's title is annoying.

Shoplifting from American Apparel
A novella about Lin's youth, partially comprised of inane
Gmail conversations.
Five words: Minimalism from a trite offender.

Richard Yates
Two main characters named Haley Joel Osment and Dakota
Fanning engage in an online affair. Osment is twenty-two
years old and Fanning is only sixteen years old.
Five words: Confusing to find in stores.

DETAILS
1. You mention that you purposely gift Lin's books to
   friends who are "less sophisticated" readers.

2. Lin once tweeted: "'pulled' my penis in opposite
   directions near its 'head' to make it 'wider' as @
   meganboyle 'poured' cocaine on the now-larger
   surface." To which his wife (@meganboyle) tweeted
   "snorted cocaine off @tao_lin's balls while grinning

as it fell 'everywhere' and then licked the balls ~6x, some of the penis ~2x," which, taken together, amounted to a decidedly more interesting piece than Lin's "How to Give a Reading on Mushrooms" for Thought Catalog.

3. His writing is listless. Lin manages to describe an emptiness short of nihilism. If you're talking to a fan you can say you "think Lin's definitely making a larger statement about technology with his particular brand of ennui."

4. Lin's metafiction style has launched a thousand shitty blog posts by amateur authors. But you can credit him for practically creating metafiction, the cousin to the independent film genre mumblecore.

How to Fake Like You've Read Mary Karr

BASICS

Memoirist and poet most famous for making her crazy Texan family famous. How crazy? Her mother gave her a copy of the book *Nausea* by Jean-Paul Sartre at age twelve.

ESSENTIALOGRAPHY

The Liars' Club

Karr's first memoir recounts the quirks and darker moments in her childhood, being raised by a manic-depressive mother and a hard-drinking father.

Five words: Wild child observes insane family.

Cherry

Karr begins to kick up her own fair amount of trouble in this memoir about her teenage years in Texas and California.

Five words: Wilder child explores life's boundaries.

Lit

Reflections on how she became an alcoholic and fought her way to sobriety through Christianity.

Five words: Wildest woman finds her way.

DETAILS

1. One of the biggest fears of any thinking memoirist is that they stand to anger and distance important people in their lives. Karr largely avoided complaints from those she wrote about in her memoirs because, as she says, she was usually "the biggest asshole" in her story.

2. The essay "Against Decoration" in her collection *Viper Rum* is one of the most succinct arguments against the overly flowery work of contemporary poets. You can say, "I think I have to agree with Mary Karr that many

contemporary poets use metaphor as an end instead
of a means to an end," whenever someone claims
someone like Rosanna Warren is the best living poet.

3. Karr dated David Foster Wallace. This is pretty much
the number one reason people who weren't already
fans of Karr because of *The Liars' Club* picked up *Lit*,
which was published after DFW's death. They met
as recovering alcoholics. Before they had even kissed,
DFW got "Mary" inside of a heart tattooed on his
arm. They are an embodiment of the metaphor "The
brightest fires burn the fastest." Within months they
had both changed their phone numbers and stopped
talking to one another. When DFW married artist
Karen Green, he added an asterisk to the tattoo and
"Karen" as a footnote below.

4. Don DeLillo, after she called him once to complain
about writing, sent her a note in the mail that read
simply, "Write or Die." She wrote back, "Write and Die."

How to Fake Like You've Read Mary Gaitskill

BASICS

Want to impress your young aunt who had a brief *Felicity*
stage during college? This writer who is best known for
being naughty is a favorite of many 1990s NYC undergrads.

ESSENTIALOGRAPHY

Two Girls, Fat and Thin

A story about two sisters, one fat and one thin (bet you didn't see that coming), that switches points of view from fat sister first person to thin sister third person (a transparent device intended to portray how the chubby one has a better handle on her world than her sister).

Five words: Ayn Rand satire, visceral, dark.

Bad Behavior

Short-story collection that contains some of the best sex scenes you'll come across in contemporary literature (the world these scenes take place in is a more sordid and morally twisted one than reality can offer).

Five words: Naughty secretary fantasy come true?

Veronica

An aged model who once walked in Paris and now cleans offices is haunted by the death of her friend Veronica and the squandered opportunities of her youth.

Five words: Remember: higher rise, higher fall.

DETAILS

1. As one of the most cutting-edge contemporary sexual fiction writers, Gaitskill is curiously afraid of one of the most sexual symbols in literature—horses.

2. In the late nineties Gaitskill championed then up-and-coming author J. T. LeRoy, a former prostitute and transvestite teen who turned out to be an elaborate ruse by a twentysomething woman. Gaitskill originally took to LeRoy likely (at least partially) because Gaitskill herself once worked as a stripper and call girl. Their stories were so similar that many presumed Gaitskill was secretly LeRoy. Around the time he was outed as an unknown struggling female writer, Gaitskill claimed that she didn't care whether LeRoy was a hoax, she thought he was interesting no matter his true backstory and identity.

3. You can shut up Ayn Rand fans by saying, "Gaitskill's *Two Girls, Fat and Thin* taught me how cultish philosophies like objectivism appeal to people who lead uneventful lives."

How to Fake Like You've Read Dave Eggers

BASICS

The man (in)famously responsible for making twee mainstream but also commendable for his various charity works and founding *McSweeney's*. Bring him up to anyone who has read a book in the last decade and they have something to say. Forewarning that anyone who claims him as their favorite author is likely quite dull.

ESSENTIALOGRAPHY

A Heartbreaking Work of Staggering Genius

A young man's quest to keep his family together after both his parents die and he is granted guardianship of his eight-year-old brother.

Five words: A heartbreaking work of self-flagellation.

You Shall Know Our Velocity

Eggers's first novel. Two friends plan out how to spend $40,000 in ways that'll make it available to random strangers.

Five words: You shall know our donkey.

What Is the What

Based on the life of a Sudanese refugee and presented as his autobiography, though Eggers fictionalized sections of it.

Five words: Point: Pain is the pain.

DETAILS

(When discussing Eggers, after every third sentence, mention how much you admire him for *McSweeney's*.)

1. People can gab about his humanitarian efforts all they want but Eggers's prior obsession with MTV's *The Real World* makes you distrustful of his holier-than-thou attitude.

2. Eggers founded 826 National, a reading program for underprivileged kids. The flagship program,

826 Valencia, is located inside of a pirate shop in
San Francisco. There's also an 826 NYC located in a
superhero shop in Brooklyn. They're best known for
attracting A-listers and not accepting your application
to volunteer. Sarah Vowell is the president of 826
NYC. Sherman Alexie is on the board of directors for
826 Seattle, which is inside of a space travel store. My
favorite 826 location is the Robot Supply and Repair
shop in Ann Arbor, Michigan.

3. Eggers's books *Zeitoun* and *What Is the What* were both
part of the Voice of Witnesses project. *What Is the What*
more likely helped you remember that "is" should be
capitalized in a title than spurred you to go and help
out in Africa.

4. At a college talk, Eggers gave out his e-mail address and
offered to respond to anyone who wrote him dismayed
by the seemingly bleak future of publishing. This e-mail
address was posted by several news outlets and blogs
poking fun at his idea of using electronic correspondence
to spread word of the rosy future of print.

5. It is necessary to misstate the book's title and use the
word "twee" at least twice while discussing the pitfalls
of *A Heartbreaking Work of Staggering Genius*.

How to Fake Like You've Read Zadie Smith

BASICS

This English practitioner of what literary critic James Wood dubbed hysterical realism tells stories about humdrum middle-class life with an exploration of the experiences and actions of often unlovable characters. Zadie Smith burst on the scene with a spectacular and often jealousy-provoking story: she received a substantial book advance at the age of twenty-one for two novels based on only eighty pages of a story. She frequently cites her love of E. M. Forster—*On Beauty* is a subtle retelling of his book *Howard's End*. She also has a penchant for Franz Kafka. A 2004 article in the *New York Times* mentioned that Smith was working on a musical about Franz Kafka with her husband, poet Nick Laird. The musical has not been mentioned by her since, but her love for Kafka remains in essays she published about his work and his biographies.

ESSENTIALOGRAPHY

White Teeth

Explores the different ways in which a culture can deconstruct the folkways and beliefs of its immigrant populations.
Five words: Deus ex machina, teeth unify.

The Autograph Man

Smith's second novel, highly anticipated but disappointing in the estimation of many, details a man distracting himself

from the loss of his father with the pursuit of a rare autograph.
Five words: Tandem in name and identity.

On Beauty
In small part a modern retelling of E. M. Forster's *Howard's End*, in large part a look at two different families living in America and making themselves mad by attempting to accept or fight against social constraints.
Five words: Hapless rendered so by happiness.

DETAILS

1. Remember to correct anyone mispronouncing "Zadie" with a short "A." She changed her name from "Sadie" to "Zadie" for the express purpose that it be pronounced with a long "A," as in "Zaydie."

2. If you're also a film buff, you can bring up Smith's piece in *The New York Review of Books* called "Generation Why?" decrying Facebook and other social networking technology after the movie *The Social Network* was released. Interestingly enough, in earlier interviews Smith herself admitted to being addicted to the Internet and wasting too much time on Facebook. Famous authors, they're just like us!

3. If your conversation partner isn't a fan of Smith, you can fall in step by pointing out the mistakes in Kiki's Floridian accent in *On Beauty* and saying it became

very distracting for you. If they are indeed a fan, you can say, "Smith creates the best kind of unlovable characters," and add that you're afraid to admit how much of yourself you see in the unstable actions of her protagonists.

How to Fake Like You've Read William Gaddis

BASICS

Feeling the need to show off? Lug *The Recognitions* around. End up with a shooting pain in your shoulder and a patronizing attitude toward plebeians whose only foray into epic novels was *Infinite Jest*. William Gaddis was an angry, bitter man, and it shows in his brilliant works.

ESSENTIALOGRAPHY

The Recognitions

At first ignored by book critics who thought it too long to be worth their time, Gaddis's most famous work outlines a painter's descent into art forgery with the message that one should lead a life filled with integrity by creating true, honest work. **Five words:** Parody *Faust*, to live deliberately.

J R

Kid genius invests single share wisely in order to make a fortune. **Five words:** Dialogue unattributed, scathing satire mastered.

A Frolic of His Own
National Book Award–winning examination of the legal system's inherent flaws and its failure to deliver true justice.
Five words: "Justice in the next world."

DETAILS

1. Gaddis was a *Harvard Lampoon* writer but dropped out of college and moved to Greenwich Village to work at *The New Yorker* (as a fact-checker, not a writer) and hang with the Beats. Now is a good time to mention some other famous *Lampoon* writers: Robert Benchley, John Updike, and Simon Rich.

2. Real Gaddis devotees read *Swallow Hard* by his daughter, Sarah Gaddis, a horrifyingly bad, bitter novel about a girl's novelist father who is more obsessed with his work than he is concerned about the well-being of his family.

3. Esme's apartment from *The Recognitions* on 23 Jones Street in New York City was the real-life apartment of the inspiration for the character, Sheri Martinelli. Suggest a pilgrimage there as a date idea if you're into a Gaddis freak.

How to Fake Like You've Read Marcel Proust

BASICS
This reclusive novelist wrote all seven volumes of his opus from the comfort of his bed. A favorite for those who have too much time to kill while in college.

ESSENTIALOGRAPHY
In Search of Lost Time (or *Remembrance of Things Past*)
Seven-volume work by Proust that details the narrator's coming of age. Characters in a state of preadulthood whose adventures in love and society seem familiar to contemporary readers despite the setting of early twentieth-century France.
Five words: Memories are luggage, madeleine episode.

DETAILS
1. Use the word "Proustian" often and use it well. His largest literary contribution is likely his establishment of the concept of involuntary memory. Involuntary memory is the feeling you get when you smell one of your mother's old sweaters or walk into your old high school. It refers to the emotions a place can conjure without even an intentional thought process on your part. Those are Proustian moments.

2. The very best move you can pull is a correct French pronunciation of *À la recherche du temps perdu*—it's "ah la

ruh-SHAIRSH due TOM pair-DUE." Also, be sure you
say "Proust" correctly. It's "proost."

3. You got through the first two books swiftly but your
   interest lagged after the drama between Swann and
   Odette was resolved.

4. A sick child who grew into a feeble adult, in his teens
   and early twenties Proust was a social gadabout; as he
   got sicker (especially after the death of his parents) he
   became more and more of a recluse. Oft mentioned
   is his bedroom, cork-lined to keep out sound. Say you
   want to "pull a Proust" and install cork next time
   you're complaining about your thin walls.

How to Fake Like You've Read Cormac McCarthy

BASICS

A real man's man, Cormac (birth name Charles) McCarthy has admitted in interviews that he doesn't "pretend to
understand women," which is why his books rarely contain them as main characters, if at all. He's also stated that
he doesn't think Proust is literature. Basically, if you panto-
mime a McCarthy fan, you can be ignorant of all other literature and still be completely in character as a cultivated
member of the literati.

ESSENTIALOGRAPHY

All the Pretty Horses
Runaway boys try to find work as cowboys, instead find corruption and time in a Mexican prison.
Five words: Lacey is a male name?

Blood Meridian, or The Evening Redness in the West
Runaway's experiences with gangs, violence, and meteor showers.
Five words: "The kid" becomes "the man."

No Country for Old Men
A drug deal sets a series of unfortunate events into action.
Five words: The movie version is better.

DETAILS

1. Cormac McCarthy unsurprisingly shared an editor with William Faulkner.

2. McCarthy never uses semicolons or quotation marks. Before officially hitting it big as a writer, McCarthy was so poor and determined not to get a regular job that he and his second wife would bathe in a lake. To fans and detractors both you can say, "He's so stubborn." Fans admire him for it; haters use it as evidence of his ineptitude.

3. Tell people they should read *Suttree*, McCarthy's novel about life as a Tennessee fisherman, before they ever pick up *Huckleberry Finn*. In fact, if they've been lucky enough to make it through their primary school education without reading *Huckleberry Finn*, they shouldn't ever read it. Twain has inspired so many other writers that you already know the story without reading the words, and most other books are better. (Sorry, I'm still bitter about the three different public school English classes that made me read *Huckleberry Finn*.) *Suttree* is one of the best examples of Twain's influence.

4. Say that when your ex-boyfriend read the ending to *The Road* it was the only time you'd ever seen him cry.

How to Fake Like You've Read Kurt Vonnegut

BASICS
Curly-haired bespectacled novelist whose wild prose aptly sums up the irony of life. The craziest people I know are the biggest fans of Vonnegut.

ESSENTIALOGRAPHY
Vonnegut once graded his own works; the grades are included in the five-word descriptions below.

Cat's Cradle

Bokonism, a made-up religion touting foot touching as worship, and ice 9, a weapon more powerful than a nuclear bomb, lead narrator Jonah on a quest to escape the island of San Lorenzo.

Five words: A+ — Man's enemy is man.

Slaughterhouse-Five

Protagonist Billy Pilgrim transcends time in order to revisit his experiences in World War II and other episodes from his life. Vonnegut himself famously lived through a horrific experience as a prisoner of war in Dresden, the same town Pilgrim is imprisoned in.

Five words: A+ — Reflections on life's fatalism.

Slapstick

In a New York of the distant future, twins try to implement a national plan to abolish loneliness as the world falls apart.

Five words: D — An examination of loneliness.

DETAILS

1. Vonnegut's chief contribution to society was his coining of the phrase "flying fuck." Without him I'd have no way to tell Jennifer Weiner that I don't give a flying fuck about her writing.

2. When speaking to fans, claim that "Harrison Bergeron" first got you interested in politics. It's a

short story about an American teenager's struggle in a society that forces people to be truly equal—as in equal in looks, athleticism, and intelligence.

3. Vonnegut's nickname in high school was "Snarf," a term for someone who sniffs girls' bicycle seats. Huh; mentally adding that to my insult dictionary now.

4. He called semicolons "transvestite hermaphrodites." Bring that phrase up next time your friend e-mails you with a litany of them.

How to Fake Like You've Read Jennifer Egan

BASICS
A novelist and journalist, this Brooklynite is a favorite among those who know better than to claim Jonathan Franzen as their contemporary favorite. It's interesting to note that she dated Steve Jobs, the Apple visionary, for a year while she was a student at the University of Pennsylvania. Plenty of her stories involve the threat of technology engrossing individuals to the point of detriment to society.

ESSENTIALOGRAPHY
Look at Me
A model deals with the aftermath of a car accident that ruined her face, making her unrecognizable to others.
Five words: Z and Internet start-ups terrify.

The Keep

A story inside a story: a prisoner writing a presumably fictional piece about a man hiring his estranged cousin to work at his hotel.

Five words: Hair box to hear dead.

A Visit from the Goon Squad

Interconnecting stories of punk rockers, rapists, and good guys, with one unifying message: "time is a goon" pushing around its unwitting victims, taking them forward with no respect to the present state of affairs.

Five words: Visual exploration of autistic mind.

DETAILS

1. Egan came under fire when, in an interview after finding out she won the Pulitzer for *A Visit from the Goon Squad*, she encouraged young female writers not to settle for writing chick lit and instead to set themselves higher standards. Jennifer Weiner, the same author who attacked Franzen for his oversaturation in the media, pounced on Egan for being callous. You can cite fear of Jennifer Weiner yelling at you as the reason you've thus far avoided Twitter.

2. Like her protagonist Sasha from *A Visit from the Goon Squad*, Egan tried to shoplift a bit as a teen, though the author herself was frightened and nervous when doing so. She said she felt jealous of the girls who could do it

without care. She also has been robbed several times before. Cheer up a friend who just had her credit card hacked by saying maybe this will inspire her to write a Pulitzer Prize–winning novel.

3. The PowerPoint chapter in *A Visit from the Goon Squad* is reminiscent of Jonathan Safran Foer's visual writing in *Extremely Loud and Incredibly Close*. "Visual writing" refers to the technique of playing images off words, side by side, in an effort to render a character's mind-set more vividly and to tell a story more fully. For example, Egan's PowerPoint chapter is from the perspective of an autistic girl. Egan is trying to show how the girl processes information in a different format from others and how this leads to gaps that others cannot close for her and connections between things that others do not perceive. Foer used visual writing by making the last pages of *Extremely Loud and Incredibly Close* unreadable with squished-together text or drawn out with dozens of periods between phrases. Foer's attempt was widely criticized, while Egan's was widely lauded. As an Egan fan, you can say that this gives you confidence that critics are capable of seeing through pure hype.

How to Fake Like You've Read Chuck Palahniuk

BASICS

One of the most vile (in theme) writers in contemporary literature—the gross-out king who hides his lack of character depth behind deeply disturbing themes of sex and violence—didn't start writing fiction until he was in his thirties (he had worked as a journalist after college).

Influenced by: Amy Hempel. He brings this fact up in practically every interview. He once wrote, "My rule about meeting people is—if I love their work, I don't want to risk seeing them fart or pick their teeth. Last year in New York, I did a reading at the Barnes & Noble in Union Square where I praised Hempel, telling the crowd that if she wrote enough I'd just stay home and read in bed all day. The next night, she appeared at my reading in the Village. I drank half a beer and we talked without passing gas. Still, I kind of hope I never see her, again."

ESSENTIALOGRAPHY

Fight Club

At first ignored, this novel turned into a cult favorite after being made into a movie. It follows the life of a bored business guy who starts a nighttime bare-knuckle fight club as a way to make his comfortable, dull life more interesting.

Five words: Support groups of all kinds.

Choke
A twisted tale of a sex addict who pays for his mom's nursing care by pretending to choke at restaurants and manipulating the unwitting Samaritans who come to his rescue.
Five words: Reach out and touch him.

Survivor
Member of a death cult hijacks a plane.
Five words: Tender lives at the end.

DETAILS

1. Palahniuk is pronounced PAU-la-nick. Most others with his Russian last name pronounce it PAH I-la-NYOOK but his grandparents decided to make it easier on American ears by pronouncing it like a mash-up of their two first names, Paula and Nick.

2. He's an active member of the Cacophony Society (Portland chapter), a group that pulls large-scale pranks that aim to challenge social norms. Think of it as a harder-drinking and more violent version of Improv Everywhere. Project Mayhem from *Fight Club* is based on the group. In his memoir *Fugitives and Refugees: A Walk in Portland, Oregon*, Palahniuk endearingly recalls his book reading at Powell's where members forced him to chug tequila and wear a Santa suit.

3. His fifth novel, *Lullaby*, was written after his father
   was murdered in a double homicide by the jealous
   ex-boyfriend of a woman whom he met in a classified
   section of a newspaper. Palahniuk wrote the novel
   in just six weeks, in an effort to come to a decision
   about whether his father's murderer should receive
   the death sentence, a decision that was left to him. If
   your conversation is drifting toward the supernatural,
   you can bring up an essay (published in *Stranger Than
   Fiction*) in which Palahniuk recounts that his mother
   dreamed his father was murdered the very night he
   was actually killed, and that his father also visited his
   sister's dreams that night, to tell her it was okay they
   had grown apart and that "the past doesn't matter
   anymore."

4. Even more extreme than the horrible death of his
   father is the death of Palahniuk's grandparents. His
   grandfather shot and killed his grandmother as his father
   watched from under the bed at age three. Afterward, his
   grandfather turned the gun on himself.

5. Palahniuk is often very careful to make sure facts he
   presents are true. To this end, the repetitive cleaning
   tips and tricks in *Survivor* are all completely accurate.
   Some of the best for your housekeeping pleasure:
   Keep bacon from curling by chilling the bacon in a
   freezer for a few minutes before cooking. A way to

keep a sharp crease in your pants is to iron them with a pressing cloth dampened with water and vinegar. Get rid of divots in carpets (from furniture, etc.) by putting ice in them (as the ice melts the divot will pop out). Clean up small shards of broken glass using a piece of bread.

Strategies to Avoid Discussing
the Major Plots Points of Any Novel

THERE'S ALWAYS SOME BOOK you've been planning to read for months, years. That one novel you bought and never got to during summer vacation—you read all the reviews about it, you even suggested it as a read for book club. Yet it sits permanently in the "to be read" queue. Suddenly, it's mentioned in conversation with someone whom you want to impress. As the well-read, well intentioned person you are, who totally means to read that book as soon as possible, you bluff. You fake it. You've read the back cover enough to have the gist of the plot down. So why not take the credit? You know I won't blame you.

But now you're stuck. You totally bluffed about reading *Let the Great World Spin* by Colum McCann, thinking that no one would challenge you on it, but someone now wants to discuss what you think was the purpose of the graffiti in the subway chapter. Let's see if we can get you out of your predicament with ten simple steps to avoid being called both a fool and a liar:

1. Mention that you read the novel many, many years ago. This allows for lapses in memory and puts your conversation partner on notice that you have a long history with esteemed novels. Say it warily to give the impression you've spent the last decade(s) reading books along the lines of Nikolai Gogol and you've been disenchanted by the bulk of them. If it is a book that came out more recently, say that you read it "when it first came out."

2. Say, "The plot seemed ostensibly dramatic—notable events occurred—but the characters were acting without real depth in the contexts of the decisions." In literature, insight into whether character development was sufficient is relative to the reader. It's a safe bet to fall back on a personal feeling that their thoughts and actions didn't provide the requisite material to presuppose their actions. Add "Well . . ." at the beginning of the statement and shake your head slightly if you want to be sure your partner gets the impression that you're beyond these banalities.

3. Do you know anything about the book? Anything at all? Or about the author? Best-case scenario, mention the one secondhand anecdote or critical thought you have in your possession and your partner will take the conversation from there in their excitement over your point (e.g.: the night he won the National Book Award,

you recall that McCann took the subway to the awards ceremony). Worst case: they won't relate to your conversation point and they'll ask you to explain what you're talking about. In that event, proceed to the next item on this list.

4. If you've read another work by the author, bring that up as something you liked more or less than the work you're currently discussing. Or bring up a movie related to one of the author's books. For example, "I recently added to my Netflix queue the amazing documentary *Man on Wire* about the real-life tightrope walker who inspired the main character in *Let the Great World Spin*."

5. If the author is current, note that they seem to be heavily influenced by Hemingway. Forewarning: you have only a 60 percent chance of nailing this point, but desperate times call for desperate measures, right?

6. If you're discussing a book of contemporary short stories, lament that it feels like an easy excuse to condense and monetize all the author's MFA work. If your partner points out the author doesn't have an MFA, shrug and say, "I figured all short-story collections were straight out of Iowa."

7. Explain away any apparent memory lapses by using one of any of the words commonly used by book critics:

"You know, this author isn't particularly indelible for me. At least, not so much as [insert a similar writer]."

8. There's a whole subgenre of books that share one common quality: they make readers want to pour a glass of whiskey or light up one unsavory substance or another. The author's somehow romantic treatment of a drink—a tidy glass of brown liquor or chilled martini—and its effect on the characters is irresistible. If you're talking about Hemingway, Bukowski, Thompson, or any of the other booze-soaked greats, mention how you had to have a scotch in hand while reading.

9. Don't attempt to discuss how the ending of the book made you feel. You don't want to accidentally claim that you felt optimistic about Ciaran's future in *Let the Great World Spin*.

10. The best way out of any discussion is to make an exit for the bathroom. If you feel like someone might be waiting for you when you come out, research the book while on the john. That's what remote access to the Internet is for. Make sure you don't take too long, and make sure to close the Web page before exiting the bathroom, lest the Wikipedia entry for "Colum McCann" pop on your screen when you decide to show your partner pictures from your last vacation.

A Gift Guide by a Bad Gift-Giver

IN NINTH GRADE I was invited to the surprise birthday of a popular guy at my school. I'm fairly sure it was by accident, that his mother picked the wrong Lauren to call from the school's phone book. I hadn't realized it was passé to bring presents to a high schooler's birthday party and it seemed inconceivable that anything other than a book should be given. Of course, I knew just the book. I had finished reading *On the Road* a couple months earlier and thought the combination of a lurid plot and simple diction would be wonderful for a sixteen-year-old boy. The fact that I mentally described the book as "wonderful" should have tipped me off that it wasn't the most ideal present. But I naïvely spent my allowance at the local bookstore and brought the wrapped book to his party.

His mother, who had called to invite me, didn't recognize me at the door. When I explained who I was, her disappointment showed in the way her eyes widened and her face contorted in a grimace-like smile. That was hint enough for me, so I apologized. "I'm sorry," I said as I stepped into her house. She just nodded as I passed, us both ignoring my awkward fate of being an unintended attendee. I'll spare you details of the dependent relationship I developed with a corner of his house that provided adequate shade from the kids who didn't care to talk to me. It's enough to say the boy was a nice person, more gracious than his friends, who snickered as he opened the present, one of the only presents brought to the party.

"Did you like the book?" I'd ask him in school days later. He told me he hadn't gotten to it yet but he'd read the back cover and it sounded great. That's exactly what he said, "It sounds great." Telling someone a book sounds "great" is the equivalent of noting the weather in an elevator. It's what one does to be polite. But I was optimistic; *He likes the book*, I thought.

"Did you like the book?" I'd ask weeks later, catching him at his locker in an empty hallway. I could feel my eyes wide, trained on him. He told me he had just been about to start it the other night. He thought he was going to start sometime that week. I was dumbly elated, not the first fool who believed a beautiful boy.

"How's the book?" I asked again some more weeks later. He responded he hadn't started it yet. At that point I realized he probably wasn't going to read it. It made me sad to think of the gift, wherever it was. Sitting alone on his dresser, probably to be handed off to some other member of his family. I pictured the mother who didn't know who I was reading it, enjoying an experience that wasn't hers to enjoy.

You would've thought I'd learn after that. Books aren't good gifts for people who don't read. But like a maniac, I keep shoving books into nonreader hands. I picture myself as not unlike John the Baptist. But instead of preaching for Jesus, I'm preaching for stories! Throwing fiction into the faces of nonbelievers, hoping to ignite a fire in their belly. I'm a selfish book-giver, choosing books with the expectation they'll mean something for our relationship, will tie us together. I'm giving you a journey. Appreciate it. Talk about it with me later.

That's the best part of having a book in common with friends: talking about it. And I find with all the books I've littered among friends throughout the years, I never know exactly when that book will catch up with us, but whether it takes weeks or years, it tends to be a great moment. I don't shoot 100 percent. I'm still only at about 40 percent for gifted or lent books being read. People don't have enough free time. They haven't had a chance yet. They have a lot of other things to read. They forgot to take it on vacation and who knows when they'll have time again. And I keep lending and giving out books, hoping to improve my batting average by hedging my bets. I'm still waiting to hear back from most of my chosen recipients. But I soldier on.

I gave a genuine guy friend *On the Road* as well. He was as gracious as the boy from the party and even texted me five years later in the middle of the night to say he had just cracked open the book and it actually was as good as I had said. I have no idea how he even got my phone number.

I've bought too many past boyfriends the book *Sophie's World*. I tell them about how it got me into philosophy and is the reason why I majored in political theory. I stupidly expect them to understand me better after reading it. So far only one man has finished it, after we broke up. There are reasons I'm single; this might be one of them. I bought Lionel Shriver's *Game Control* for a man I thought wanted to change the world after a couple great, inspiring dates. We broke up before he cracked the spine. *Super Sad True Love Story* went out to a technologist who managed to create the same event as the title in my life without ever reading the book (or returning it). I fell in love with a man who enjoyed *Middlesex* as much as I did. Along with another who articulated the same reasons I felt *A Farewell to Arms* was overrated. And a third who borrowed *Choke* and gave it back with sarcastic annotations.

I lent out a copy of *The Bad Girl* to a guy who made me laugh. I lent *Gourmet Rhapsody* by Muriel Barbery to a coworker who hasn't mentioned it since. My copy of *Rules of Civility* has gone out to two friends who couldn't stop mentioning it, although one of them hasn't given it back yet. I got in a fight with one of my closest friends after I lent her *The Sound and the Fury*, and she returned to me a different edition of the book, having lost mine. My upstairs neighbors received *Richard Yates* by Tao Lin and Michael Chabon's *The Amazing Adventures of Kavalier and Clay*, and I, possibly drunkenly, threw *Family Circle* by Susan Braudy at them too.

Who knows, I can't find it. After too much wine or beer at my house I've run to my bookshelf and thrown everything from Jonathan Safran Foer's *Extremely Loud and Incredibly Close* to Thomas Bernhard's *The Loser* at people. Parties at my place mean you're more likely to leave with books in hand than drugs. Sometimes the books return, sometimes they're out of sight forever.

I present this all as a cautionary tale to my following gift suggestions. They might be a bit off pitch and you may not ever hear from the recipients about the books again, but if you do nail it, I expect a commission.

Gift Guide

OBNOXIOUS MOTHER-IN-LAW
The Puppy Diaries by Jill Abramson. Frilly but with an author and contributors who pack enough notable punch that she won't feel her intelligence is being insulted.

INDIFFERENT FATHER-IN-LAW
Anthony Bourdain's *No Reservations*. No fuss, no sentiment, no problem.

MOTHER WHO WON'T ADMIT TO READING
A political biography, but avoid the latest presidential biography; she has too many opinions on him. Veer toward the more nostalgia-inducing figures, anyone from the seventies except Henry Kissinger.

FATHER WHO DOESN'T READ

Guy Fieri's *Diners, Drive-ins, and Dives: An All-American Road Trip . . . with Recipes!* All fathers who don't read like Guy Fieri.

SLATTERNLY SISTER

After the intensity of *The Virgin Suicides*, I tend to think of Jeffrey Eugenides as the patron saint of lost causes. Not for his ability to remedy futility, but for his ability to capture it. Present it and hope your sister learns something from the Lisbon girls.

BROTHER WHOSE FAVORITE BOOK IS ANY HARRY POTTER

Are you trying to better him? Try *The Princess Bride*. If you're just trying to please him, go for *The Hunger Games*. Do you hate your brother? Gift *Twilight* to his e-reader; the discretion and accessibility guarantee he'll secretly read it, kind of like it, and feel shame because of it.

KIND OF STILL SLUTTY BEST FRIEND FROM THE SORORITY

Why Men Love Bitches.

BOYFRIEND WHO MAY VERY SOON BE AN EX-BOYFRIEND, BUT IT'S THE HOLIDAYS, HEY

Give him something you want to read and if he doesn't immediately take to it, ask if you can borrow it to read, very quickly, and say you'll give it right back.

BOYFRIEND WHO ISN'T AWFUL

The Portable Dorothy Parker, because he's not awful and there aren't that many men around who are so great they'll read Dorothy Parker and laugh about it with you. If he seems uninterested in the female satirist, he is awful.

HUSBAND OR WIFE

I don't know—you *married* them. Do you have to get them presents anymore?

GIRLFRIEND-LIKE FIGURE IN YOUR LIFE

Test her! Throw her to the wolves. Give her some good Russian literature and pat her on the bottom as you send her out the door. If she returns having read it, maybe you keep her.

REAL GIRLFRIEND

Show her you love her! Set her up nice with a full collection of something wicked, like Oscar Wilde.

How to Speak Condescendingly About the Most Revered Authors/Literary Works

FOR THOSE TIMES WHEN you're looking to cut someone down to size, the following approaches will shut the yapping maw of any pompous beast dropping names of the literary works and artists they seem to feel so intimately acquainted with. Here's a secret about readers: we don't need to fawn over our favorite book. Hit the right button and you'll get oozing fandom, sure—but we have nothing to prove. The couple at the dinner party who raise their voices debating whether or not Nick Carraway is attracted to Gatsby? They don't know what they're talking about. Beat them at their own game by regurgitating one of these lines in a wry, disenchanted tone of voice.

- "I don't dispute *To Kill a Mockingbird*'s place in American literature as the middle-school conduit from Nancy Drew and the Hardy Boys to the more complex world of social novels; it's just that if she were really talented, Harper Lee would have felt compelled to write another book. I don't believe *Mockingbird* is good enough to be both a debut novel and a magnum opus." —*To Kill a Mockingbird*, Harper Lee

- "People too often focus on the mythic quality of the story when the real beauty of the book is in the characters. Readers get hung up by the way the story line is represented in today's popular media. It's not about a whale." —*Moby-Dick*, Herman Melville

- "It's not that *In Search of Lost Time* didn't hold my attention, but I kept getting distracted by all the details that I knew were biographically in line with Proust's life and wondering how many of the 'fictional' details actually were drawn from his life, just undiscovered by biographers. I spent more time Googling his acquaintances than reading the book. Knowing that a work is heavily based off an author's life makes me distracted by the possibilities of which details are drawn from real life." —*In Search of Lost Time*, Marcel Proust

- "Sure, *Nineteen Eighty-four* was good when I was in high school and felt like the bells between classes and announcements over the PA system controlled my life; however, one cannot doubt that Orwell oversensationalized the individual's loss of control in society. Now, Orwell cannot stand up to Huxley since the rise of the Internet. Huxley won the battle for the future with his society of inescapable individualism. Although *Brave New World* contained too many overt metaphors—and the repetitive nonsense with the 'ford' joke was irritating. His satire could benefit from a little more subtlety." —*1984*, George Orwell

- "Whenever I hear someone lavishing praise on *The Great Gatsby* I assume they have subconscious desires to live a life like Gatsby. Oh, I don't think that about you, though." —*The Great Gatsby*, F. Scott Fitzgerald

- "Really? You enjoyed *American Psycho*? I never got around to reading it; I usually don't share the same interest in books with my brother, who has it on his bookshelf next to his limited-edition *Scarface* DVD." —*American Psycho*, Bret Easton Ellis

- "George Eliot captures the dynamics of nineteenth-century politics, to be sure. However, I can't stand her exasperating commentary on every action of the characters. Do I really need to read pages on the lingering silences in Dorothea's interactions with her husband? She'd be right to apply some of Hemingway's iceberg theory and let the reader uncover her intentions for themselves." —*Middlemarch*, George Eliot

- "I cannot get past the sense that the stories would have fared better if they were presented as self-contained pieces in a collection, rather than being forced to cohere as a novel. As in Byatt's earlier novel, *Possession*, the structure doesn't feel organic. I have to assume she thinks she's using some sort of thematic tool, but the result is just long-winded and rambling." —*The Children's Book*, A. S. Byatt

- "The superficiality of the characters makes it hard to stay invested in their fates, and I cared less and less about their philosophical inclinations as the conversations became more and more contrived." —*Women in Love*, D. H. Lawrence

- "*The Duel* perfectly captured gossip bred from the monotony of small-town characters. It falls flat, however, when Chekhov takes moral positions and veers into righteous Tolstoyan hyperbole." —*The Duel*, Anton Chekhov

- "I wasn't confused by the overabundance of pronouns, I was just uninterested in watching Mantel flex her historical knowledge through arbitrary references instead of trying to create a real story or venturing substantive commentary." —*Wolf Hall*, Hilary Mantel

- "Seven hundred pages of plotless onanism and stylistic indulgence." —*Ulysses*, James Joyce

The Written Word

THERE'S A REASON BOOK lovers are the last ones to hold out in this digital revolution. Music and movie lovers have never had the same pleasure that book lovers have in being able to identify on sight a fellow fan of Tolstoy or Didion. What does the e-reader revolution mean for all of us who get a thrill from noting the book in a stranger's hands?

Music devotees meet other fans at concerts or by recognizing concert T-shirts. Movie lovers will wait in line together at midnight for the first peek at the new Tarantino film. Readers, however, have it better; they carry around the objects of their worship and roots of their collective bond everywhere they go. I saw a girl reading Chuck Palahniuk's *Pygmy* on the subway the other day. Her mouth was pressed tight as if she were trying to prevent herself from laughing in public. It made me think of when I read Palahniuk's *Choke* in ninth grade and learned things about sex that I had no idea humans could do and how I had to tell my friends about every disgusting scene in the book. Or how his book *Survivor*, about a doomed plane ride, haunts me whenever I board an airplane. That's the kind of connection we can form with a stranger—even without their knowing—just by seeing a particular book in their hands. Life happens alongside the act of reading—a story is forever mixed with where we were and what we were doing while we were reading that book. To see someone else reading that book is to know that you share a sort of intimate experience. Ten years from now the girl from the subway will be in a restaurant and find herself next to a man reading Palahniuk, and she'll be transported back to her first job, her commutes home, and the memory of trying her hardest not to laugh out loud while reading the clipped absurdism of *Pygmy* on the train.

Except in ten years, print books themselves may be a thing of the past. I fear as digital books become ubiquitous, the tradition of reading may remain as strong but the ability to sight fellow minds will be disintegrated. As book covers slip from hands and are replaced by plastic tablets, readers lose the wonderful, clandestine opportunity to quickly create a mutual understanding with strangers. Then what will we be left with? And what about other print traditions? If bookstores vanish, where will an author's book readings occur? And book signings? What will authors sign?

I might be biased. One of my most cherished possessions is a book signed by Jeffrey Eugenides, dedicated "To Lauren: the note margin girl," because I had marked up his *The Virgin Suicides* with notes in a fervor to discover why the virgins killed themselves, like it was a mystery to be solved. My bookshelf is neatly organized by genre. I've been known to go back and buy the tangible version of a book I've read on my iPad just to have it on my shelf.

Any reader will tell you fondly that it's nice to be able to fit six books into their iPad instead of trying to cram them into a suitcase when they go on vacation. It's nice to buy another book by the author you just finished reading at midnight on a Friday. It's wonderful to be able to flip the e-pages with one hand, so the other can stay snug under your pillow while you drift to sleep. And certainly it's a nice plus that e-books are cheaper than print editions (but not cheaper enough; surely we can do a little better than one dollar or two off an electronic version of a book). The ease of having all those letters and pages in one slim tablet isn't anything to disregard as well. I would have loved to read *A Man in Full* by Tom Wolfe on the iPad instead of lugging that thing around.

But how can I scribble in the margins? E-readers have been trying to tackle this with several programs to help users annotate, but they seem too purposeful, too controlled, too able to be indexed. The joy in note taking is the ability to scribble wherever without explicitly committing to drawing some form of conclusion or highlighting a point. I want to mark up a book and forget about it. I want to never see that note again until I'm trying to remember the name of that one character's daughter and I open the book and there's that phrase that so concisely summed up where I thought my life was at the time I was reading the book.

Back to my most lamentable point: we're losing the ability to recognize fellow readers. You can bring your Kindle into a coffee shop, but unless I position myself right behind you, I have no idea what you're reading. Will we wear T-shirts to ensure we can identify one another? Will I have Franzen poised at a desk emblazoned on my chest the same way a teenager dons Lady Gaga in front of a mic? Will we place promotional stickers of books on the outside of Kindles so everyone passing by knows what that tablet contains? How sad this future seems, with no latest best-seller in the hand of the person next to you on a plane for you to strike up conversation about. You'll have to wait to meet fellow fans until the author comes to town.

PART IV

Snark Bait

Twitter-Sized Reviews of Memoirs

MEMOIRS SEEM TO BE a target for pithy reviews by critics. And let's face it, plenty of them deserve it. Memoirs bother me because they're too easy. Instead of worlds created out of imagination, they're descriptions of one's own life, and plenty of those lives are a bit too boring to have deserved a book deal. They tend to consist more of navel-gazing than of action. Memoirists are the stars of their own story, even when they are writing about their famous sibling. Sometimes they have astounding true tales that are worth sharing (looking at you, Jeannette Walls) but often they're just a person riding the wave of a burst of popularity or exaggerating life experiences to make them seem exotic (hi, Snooki).

If alien life forms were to come to Earth and examine the memoir section for an idea of what it is like to be human, they'd see five possible roads to take in life: fifteen minutes of fame, misery, celebrity, literary, and family member of celebrity.

The memoirs by those experiencing their fifteen minutes of fame don't have to be written well; they have enough hype from CNN and other media outlets. The misery memoir usually begins with the author's awful childhood of mean mommies and daddies and not a lot of money. Celebrity memoirs are often ghostwritten and not too revealing but always released first in hardcover with a beautiful face on the front. Memoirs by literary figures regularly straddle the line between misery memoir and celebrity memoir, depending on how good of a writer they are; the more miserable, the better. My personal favorite type of memoir is the kind written by a (nonfamous) family member of a celebrity. These start out the same way, every time: watching their beautiful and famous mother put on makeup, in the audience of their Casanova father's performance, or switching on the television to catch news of their famous daughter's latest scandal. The author crams more gossip in them than you'll find in any of the other categories; they often don't have allegiances to the famous friends who cameo in their relative's life and need to create waves to boost their sales. What better example of this than *Mommie Dearest*?

Here are critiques of memoirs that are as quick as I can possibly make them. If you're going to package and sell a book of your life without having a full enough life to warrant a narrative, I'm going to give you as little of my time to review it as possible:

- "Annoying blond woman harps about her extravagant vacation and upper-middle-class premenopausal problems for four hundred pages." —*Eat, Pray, Love*, Elizabeth Gilbert

- "Young New York Jewish girl has same boring life experiences as every other young New York Jewish girl, gets book deal." —*I Was Told There'd Be Cake*, Sloane Crosley

- "Teaches you the ways to seem super narcissistic and annoying to clients. Also, how to exploit international economies!" —*The 4-Hour Workweek*, Tim Ferriss

- "Should've been titled *I Enabled My Son to Become a Drug Addict and I'm Getting Rich Because of It*." —*Beautiful Boy*, David Sheff

- "Never mind." —*A Million Little Pieces*, James Frey

- "No, really, I loved my dad." —*High on Arrival*, Mackenzie Phillips

- "How to make growing up destitute with completely insane and irresponsible parents kind of seem cool." —*The Glass Castle*, Jeannette Walls

- "Your eyes may never roll back to the front of your face after reading this book." —*And the Heart Says Whatever*, Emily Gould

- "Oh look, he found a way to get a blow job without having an intern do it." —*My Life*, Bill Clinton

- "Is this really by the same person who wrote *The Liars' Club*? You sure? Because that woman was interesting." —*Cherry*, Mary Karr

- "OMG OUR PRESIDENT TOTALLY DID BLOW A FEW TIMES IN HIS LIFE!" —*Dreams from My Father*, Barack Obama

- "I didn't read the book, I just think Ayaan is super hot." —*Infidel*, Ayaan Hirsi Ali

- "Handbook on how to get an eating disorder and maintain it, marketed to adolescents." —*Wasted*, Marya Hornbacher

- "A valid case for why celebrities should never write anything, ever." —*Storitelling*, Tori Spelling

- "Sensationalism at its best." —*Bad Mother*, Ayelet Waldman

- "Barbara Sinatra was such a bitch." —*My Father's Daughter*, Tina Sinatra

- "There is no way I can make this joke without using the phrase 'wire hangers.'" —*Mommie Dearest*, Christina Crawford

Book Critic's Bag of Tricks

USE THIS HANDY GUIDE like a cheat dictionary for how to make your sentences seem smarter. These words often appear in conversation when someone is aggrandizing or debasing whatever is on people's radar. Consider this your SAT vocabulary-word cheat sheet, where "SAT" stands for "Snarky Author Types" instead of "Scholastic Assessment Test." For the love of God, please use these words sparingly.

CULTIVATED
Refined; required in the *New York Times* Style section.

> *The intelligentsia working the shop cultivated local flavor with their organic, homemade kombucha and an interactive Tumblr account.*

MOROSE
Sullen; hipster-style depressing.

The dark, plaid wallpaper and antlers perched above the
fireplace give the Brooklyn bar a morose atmosphere.

COMPELLING
Interesting; for when you've already used "interesting" in
the sentence.

It was interesting to see the congregation of authors speak
on the topic of self-promotion via social media tools at the
book festival panel, but the most compelling aspect was the
denunciation of Twitter as a spam portal for your latest
book reading by the moderator.

INEFFABLE
Indescribable; a food reviewer's favorite crutch.

I thoroughly enjoyed the moules marinée's hearty, dense
garlicky flavor, but the lamb chops with pistachio tapenade
were ineffable!

UBIQUITOUS
Everywhere; kaffiyeh scarf in winter 2009, *The Girl Who Kicked*
the Hornet's Nest circa summer 2010, hot toddies at the start of
winter, unemployed actresses in any Los Angeles It spot.

The ascent of Charlie Sheen into a ubiquitous meme rendered him played out in the media within weeks.

PITHY
Snarky; Gawker, years ago.

Her pithy comments while reviewing Sarah Vowell's latest work earned her the ire of NPR fanatics.

UNTENABLE
Indefensible; for the girl who went to Harvard and doesn't want to directly admit her employer fired her

The situation at dinner became completely untenable when his droll date announced that she thought The Book of Mormon *was unfunny.*

SUPINE
Lying face-upward; the position of models in American Apparel advertisements.

She thought her supine position on the bed made her look sexy; he thought it made her look drunk and listless.

INDELIBLE

Unforgettable; Nicki Minaj's pink dildo accessory at concerts.

> The indelible impression left by Jonathan Tropper's The
> Book of Joe made her positive that she'd never agree to her
> agent's request that she write a memoir.

FRISSON

Excitement; what you don't experience when you find out
the girl/guy you like owns a bichon frise dog.

> Their frisson at the idea of starting their own literary blog
> was not tempered by the group's previously unsuccessful
> attempts at getting published; in fact, it fueled their ambi-
> tion.

DIDACTIC

Intended to teach; your high school boyfriend's father. The
quality of a person you don't want to bang.

> Ryan's didactic father always took the five minutes before
> we headed out to teach me some mundane fact about the
> movie's historical setting or the restaurant's marinating
> process.

AUTODIDACT

A self-taught person; the person you should've banged in college but they probably didn't stick around for long.

He played up the idea of himself as an autodidact by stress-ing how he taught himself computer development and French in order to land that job in Paris.

LUVVIE
Ebullient actress; Bette Midler.

She didn't mind that they called her a luvvie in reviews of the Off-Broadway play; she was getting plenty of requests to appear at drag clubs around the city.

DIGRESS
To go off-topic; what happens too quickly whenever one is trying to have a serious conversation while drunk.

It was hard to get a read on how much she actually knew about the plot of the novel; she kept digressing into who was playing whom in the movie version.

DEUS EX MACHINA
An unexpected twist that provides a contrived solution to a problem; the device that ended the *Dynasty* television series.

I'm not sure if I believe her deus ex machina of a taxi that arrived at just the right moment to save her from having to sleep at his place in Jersey.

PORTMANTEAU

Melding of two words: "jeggings," "gaydar," "frenemy."

I thought the portmanteau of "soundscape" was pretty self-explanatory, but he seemed uninterested in my suggestion once I explained that we'll listen to music to create the atmosphere of Paris on our staycation instead of actually visiting it.

ACERBIC

Sharp; the best way to hear someone's wit described.

Her acerbic comment about the event's being "a grandiose celebration of not being very grand" made me spit out my coffee with a chuckle.

ENNUI

Boredom; the attitude of the main characters in an independent French film.

The ennui of the group made it clear they were uninterested in participating in his plan to start an ironic Avril Lavigne fan site.

OEUVRE

Complete collection; *West Wing* boxed set.

Due to the author's oeuvre it was surprising when it was announced that his next work would be science fiction.

HAM-FISTED

Clumsy; your attempts at writing poetry.

His ham-fisted arguments rendered him more incoherent than Glenn Beck.

ESOTERIC

Understood by only a small group of people; the ending to *Lost*.

I found the lecture to be fairly esoteric, specifically when he started referring to the motivations of individual Star Wars characters as they applied to our subject.

INEXPLICABLE

Unable to be explained; Kardashian fame.

Her inexplicable arrival at the bar made the others wonder whether she had been watching their Foursquare check-ins.

LIMPID

Clear; *Friday Night Lights'* eyes.

He spoke in limpid tones so there was no misunderstanding about the fact that this would be the only time he'd let them crash on his couch.

PEJORATIVE
Belittling; how Mel Gibson treats everyone.

She found it pejorative when her friends giggled at the news that she had obtained a job as a social media manager.

LACONIC
Concise; text messages to that ex-boyfriend/girlfriend.

His breath was so awful it was a relief that he tended to be laconic whenever he'd talk to her at work.

Give It to Me Cheap

I NEVER CARE WHAT a book looks like. In fact, I'm more likely to scorn the overpriced and overdone hardcovers in favor of a flimsy paperback. When confronted with a slew of editions of the same book, I'll go with the ugliest and cheapest, feeling sympathy pangs for the odd one out. Crappy paperbacks are tributes to use. They allow for cracking the spine and folding pages and rolling the book into a purse or shoving it into a cramped airplane seat pocket. Hardcovers always feel like a weighty, priccy possession. Something to be cared for, with a dust jacket to keep on the book and sometimes a small, overly delicate ribbon attached to the binding to be used as a bookmark. The day I change my reading habits to preserve the appearance of my books is the day I start to die inside, for surely I'll have stopped loving to read. My relationship is with the stories; the book is merely the *portal that must be able to meet whatever obstacle* comes our way so I can comfortably proceed with the story.

When I close a book after finishing, I like when it's unable to rest tidily together, its edges furled out from being open for so long. The cheap books are the ones that get me. Tissue-thin paper begs to be abused. I fold over five pages at a time to mark my one spot. I flop the book open at the center and hear the creak of the binding as I make the back and front covers touch. Sure the font in this five-hundred-page book is only eight points and the pages are so thin I can see the type on the other side, but a real reader doesn't wait and save up for an expensive printing of a book. No, the world inside the book transcends the minor inconvenience of transparent pages and a visit to the optometrist afterward. By the end, the book has been used up, but I'm of the opinion that a good book should wear its readability, should bear the remnants of when the owner turned the page too quickly or couldn't put the book away while cooking a meal. I like to revisit it and find pages full of life from when I was reading the book. Annotations, dog-eared pages, coffee stains, and pasta sauce splashes—the only torture I won't put a book through is tearing out a page; the thought of it, missing a couple paragraphs of the narrative or more, gives me deep anxiety.

As for book covers, there is something to be said for covers that present nothing more to you than the title and the author. Receiving the Back Bay editions of J. D. Salinger, with their only flourish being the calligraphic titles, made you feel the potent, controversial nature of reading the book. The understated goes far when you're holding a book

that speaks for itself. I wouldn't mind owning a whole collection of books that were adorned with nothing more than
the title and author.

I'm aware there are some book covers out there that can
double as art but I'm unable to find the interest to invest in
them. For those who do, I recognize the magic in a cover
that strongly brings forth the message of the book. There
are artists, like Chip Kidd and John Gall, who have created
careers out of interpreting stories in beautiful forms on book
covers. Great covers are not a graphical summary of story;
they're the artist's comment on the message, their interpretation of the images inside. Take, for example, the brightly
hued cover by David Pelham for Anthony Burgess's A Clock
work Orange, with the protagonist Alex's countenance completely featureless, except for one wide eye, set against an
empty backdrop. After reading through the book, the featureless face and barren background along with the garish
colors highlight the dystopian mind experiments done on
the incorrigible teenager.

Give me books cheap and dirty. I won't say no to plain or
unadorned books because I'll feel less guilty when using and
abusing them. Books shouldn't come with accessories like
dust jackets, embossed lettering, belly bands, foil stamps,
and the like; the reader should value the story, not an expensive cover. Happiness is a bent page.

How to Succeed in Classifying Fiction Without Really Trying

WHEN IT COMES TO determining the genre of a book, there are two things to keep in mind. First, it doesn't matter. The only advantage to identifying which movement a work belongs to is you can anticipate the possible themes and techniques in that book. It has the appearance of being the shortest distance to determining what you want to read. It makes the job easier for the marketing departments of publishing houses and sales assistants at bookstores. In the mood for some repressed sexuality? Pick up some Romantic authors. Want to get tossed around with fragmented thoughts? Grab a seat and a modernist. You love some blood and guts mixed with the prosaic? Stick with transgressive fiction.

This advantage is dampened by the second point—that someone somewhere disagrees with the stated genre of the book. For every mention of *Ulysses* as the pinnacle example of modernism, there's a scholar crying out that it's actually a sort of pre-postmodernism distinct from both modernism and postmodernism. Movements in fiction are fluid; there has never been a work isolated from both its predecessors and its successors. Due to this, classification is a wholly subjective event; the pendulum can swing greatly, with nuances seen by some readers and not at all by others. Is David Foster Wallace to be classified as a post-postmodernist because he employs a metafiction narrative at points where postmodern writers highlight reality? Do we follow James Wood's prompting and call Zadie Smith an example of hysterical realism because of the desperateness in the daily life of her characters, or are her themes of futility solidly postmodernism? These questions cannot be answered.

So, classifying fiction is a nebulous process, the outcome of which will never be accepted by all. Why bother? I have two points to argue for that as well. First, understanding the movements leads to a better understanding of literature. The fact that movements seemingly spill into each other, each influencing the next, gives us otherwise inaccessible context around the author's motivations with theme and technique. One can realize George Eliot may have bucked the common Romanticism of the time in favor of realism so she could more adequately address her politically driven views. One has the ability to predict that a surrealist work might contain themes of reason being usurped by love. Second, the marketing departments are onto something: the attempt to classify does, for the most part, make the gist of the book more accessible and thus make it easier for readers to determine which books to buy. Stark differences do exist between the genres, even if they are all leaking at the seams and there is no consensus on the exact components of each categorization.

Due to all this, I broke down the genres as simply as possible in case you ever find yourself browsing the demarcated sections of a bookstore. In the guide below, I've included authors who tend to embody the genre, a description as short as I can make it, and, for extra context, an example of a movie containing many of the same elements as the style.

Transgressive

Notable authors: Chuck Palahniuk, Dennis Cooper

Designed to shock the reader with hyperdetailed descriptions of sex and violence. The protagonist leads a normal yet empty life that he (it tends to be a he) is able to transcend only by acting out in intensely graphic ways.

Movie version: *Serial Mom*

Modernism

Notable authors: James Joyce, D. H. Lawrence, Joseph Conrad

Often first-person work focused on the mundane and employing a pessimistic viewpoint, colored by an anti-Romantic sentiment and a dedication to individualism.

Movie version: *Wild Strawberries*

Realism

Notable authors: Stephen Crane, George Eliot

Often interchangeable with the genre naturalism, realism depicts moment-to-moment reality without superfluous symbolism.

Movie version: *Before Sunrise*

Romanticism

Notable authors: Mary Shelley, Nathaniel Hawthorne

Exalts feelings and intuition over science and reason. For example, Hawthorne's *The Birthmark* is a story about how a husband's quest to remove his wife's birthmark through scientific experiments eventually kills her.

Movie version: *Gattaca*

Surrealism

Notable authors: André Breton, William S. Burroughs

Utilizing unexpected juxtapositions to force readers to think beyond logic. An easy way to remember how surrealism combines two seemingly opposite aspects is to remember that André Breton, one of the biggest proponents of surrealism, published *Anthology of Black Humor*. Black humor did not exist as a phrase at the time; Breton created it as a way of describing humorous handling of taboo topics.

Movie version: *Being John Malkovich*

Postmodernism

Notable authors: William Gaddis, Don DeLillo

Departs from modernism by employing plural realities. A classic example would be how Joseph Heller consistently has the protagonist of *Catch-22*, Yossarian, remembering the death of Snowden through a series of flashbacks that occur throughout the novel before coming to a point during the actual event.

Movie version: *The Butterfly Effect*

Magical Realism

Notable authors: Isabel Allende, Gabriel García Márquez, Toni Morrison

Works containing highly lyrical evocations of hybrid worlds mixing reality and fantasy.

Movie version: *Pan's Labyrinth*

Southern Gothic

Notable authors: William Faulkner, Barry Hannah

Set exclusively in the American South, with very ironic and often morbid events.

Movie version: *Seraphim Falls*

Hysterical Realism

Notable authors: Zadie Smith (ask James Wood)

A listing of the mundane with such precision that it becomes plot.

Movie version: *Clerks*

Metafiction

Notable authors: Italo Calvino, William Goldman

Fiction in which the fact that it is fiction is addressed, usually as a device to tell the story of writing the book.

Movie version: *Synecdoche, New York*

The Literati

---

or, Why Ernest Hemingway Once Told John Updike Literary New York Is a Bottle Full of Tapeworms Trying to Feed on Each Other.

FOR AS LONG AS you can remember, you've wanted to be a writer.

Fantasies of lounging in some dimly lit and quiet bar with men in plaid suit jackets having conversations about Proustian memories over bourbon and ice filled your mind while you sat under fluorescent lights in your high school math class.

You arrive in the big, bright city with ambitions to live in squalor, by the light of your computer. You arrive hungry to debate the nuances between Atwood's characters in different works, Lin's use of quotation marks, and your own hopeful plot twists. This is the place you've dreamed of for debate, for discovery, for showing off the obscure tidbits of literary trivia you've absorbed through years of study. You are in New York City.

Suitcase in hand. Eyes on the sky. You can't afford to

live in the city so you move to Brooklyn or Queens, or you manage to find a place in the city and you have three other roommates in an apartment designed for one (but that's okay because everyone you're going to learn from is doing the same, and this will tighten the bond with your new guides, the other unpublished writers).

And you say, "I have this story," or "I want to write for this place," and the others smile with the knowledge of how hard you, young one, are going to get broken. "Sure, I've applied to those too; I've sent stories around as well. Good luck," your roommate says while getting ready to go out to a bar they can't afford. But you think you're different. You've got the real fire, the inexhaustible flame, not the kind that dissipates to ennui so harsh that your peers can't find time in their languishing days to write. It's the red-hot, bursting-out-of-your-skin, causing-you-to-smile-on-the-street-with-sudden-inspiration sort of fire of ambition. The perfect opening to a chapter hits you as you're waiting for the subway, and you take the long walk home instead because you've been hit by too much energy to sit down.

You pick like-minded people to surround yourself with, whether it's through a living arrangement or friends you make at work or people you were introduced to through hometown acquaintances. *Why are you here?* "I want to be a writer"—you've already messed up by exposing this vulnerability. *Oh, what kind?* "I have a novel I want to sell." *Do you have an agent?* "No." *I might know someone. I know a lot of people in publishing. By the way, have you read such-and-*

such new book? "Yes! It was great!" *I thought it was overly droll.* "Oh." *I heard so-and-so got a book deal.* "Oh! Good for them!" *My friends who work at so-and-so's publisher say the editor wants to kill so-and-so—she's an awful writer. It's going to be such an atrocious book.* "Well, what's it about?" *It's a memoir. She got the book deal from her blog.* "Oh. Huh." *Did you really think such-and-such book was great?* "No, I mean, I guess I just liked the point of view." *Oh God, that was such a gimmick. Seriously. So-and-so couldn't resist being kitschy if his life depended on it.* "Oh, sure. But I think it serves his purpose well. It's an untrustworthy narrator; that's why it might strike you as unauthentic." *Please! We're going to have to teach you. There are, like, only two good authors, and that's [insert completely esoteric author] and [insert completely obtuse author].* "Yeah, I mean, I have read a couple of so-and-so and so-and-so's works . . ."

And then you step and repeat that conversation enough times until everyone in the city has worn you down enough that you can't possibly think anyone has any talent. And you grovel in the bottom of the bottom of the bottom of submissions, sending in story after story as your hatred for the author who got a book deal for his seemingly boring and stagnant and as-of-yet uneventful life just because he had a lot of hits on his blog for a post that was nothing more than a superficial imitation of Wallace's metamodernism grows and grows until it scabs your brain, specifically where your ambition center is located.

And the conversations you expected about the humor in

Kafka are overshadowed by deriding Lorrie Moore's novel or talking about how someone who works for the publishing house that Jonathan Safran Foer is published by said his next work is basically nothing more than commercial fiction, and you start to realize that literature is a social event for these people, a way of defining themselves and a way of living but not a purpose. And this realization makes your hands seize up next time you're about to put story to paper because you write to avoid this way the world works, not to try your hand at entering this contrived pecking order.

Until finally, one day, you give up.

MY SOLUTION FOR ALL the young writers being discouraged to the point of giving up is simple.

Murder the others.

Poison their overpriced vodka and soda while they're in the bathroom.

Shoot them in the face when they're asleep.

I'm talking about the people who read only to criticize and who talk only to condescend. Rid our planet of them. Write your story. Send it in to one thousand publishing companies and when you've received enough rejection letters that you could paper your walls with them, send it out to five thousand more. Carry a stiletto knife so any person who comes within five feet of you and isn't shouting out, "YOU MIGHT BE THE BIG BRIGHT LIGHT IN THE DIM WORLD OF AMERICAN CONTEMPORARY LITERA-

TURE!" can get stabbed in the gut. Keep cast-iron frying pans in your kitchen so you can invite your date in for a nightcap, then whack him in the head when he's not looking if he tells you anything less than, "YOU MIGHT WRITE THE WORDS THAT SAVE US ALL."

To Be Read

MY GRANDPARENTS WERE THE first people I knew who *really* read. In the open, glasses on, with a reading lamp lit up next to them. They had the extravagance of several bookshelves, one in the basement for business books written by men like Lee Iacocca and classics long abandoned but still beloved, like Rudyard Kipling. In their living room was a constant, revolving stack of books around a small side table; those were the books they were currently reading or about to read. Then they had a china cabinet for books on display (favorite, notable, and emotionally significant ones) or books recently finished but yet to be carried down to the basement. They'd sit and do the crossword in the morning, my grandmother arguing with my grandfather because he used a pen on it. They were the parents I desperately wanted, as so many pairs of grandparents are for kids.

My mother spent most of her life avoiding intellect. Jumping on a plane to Italy before her high school graduation, skipping college. When we played trivia games around the house, we couldn't ask her outright for the answers; if we tried she'd give us a patented sigh and eye-roll. It's only when my siblings, my father, and I started debating answers with each other that she'd step in, roll her eyes again, and tell us that the prime minister of Israel is Benjamin Netanyahu, glucose is a monosaccharide, butter in French is "beurre," and "Somethin' Stupid" was the name of Nancy Sinatra's duet with her father. I've never seen someone who so thought knowledge was useless know so much. The first time I heard "autodidact," it was like I finally had a word for "mother." My mother only appreciated learning from experiences. In her mind, you figure out the producer of a record because you go to the concert. You tour the Grand Canyon to find out how large it is. You try out mountain climbing to learn the names of the equipment. She read voraciously, but no one ever saw it; she'd surreptitiously sneak a book into her bedroom at night. Her reading consisted almost entirely of biographies. It's from her that I get my love for memoirs of the relatives of famous people. Fiction was a luxury she didn't have time for. Learning how Esther Williams met her second husband is important to understanding the world. She could dominate on crosswords, but it appeared to actually pain her to give me the answers. She would say with a sigh, "I'm no good at that," when I held a crossword and pencil toward her. So I'd find ways around it: "There's this really great kind of Japanese beer, hmm . . . I can't seem to think of the name," I'd say absentmindedly, hiding my crossword under the table. She'd respond, "Asahi probably, and don't drink beer, Lauren. It's unladylike." I had my answer.

My father decided to never bother to read a book, since I had such a stranglehold on the market. My book reading was something he was proud of, his pride akin to the feeling one might have toward someone being courageous in the face of adversity. To him, my reading habits made me similar to a cancer patient who doesn't cry during chemotherapy, an amputee who learns how to run. When feeling particularly pleased with me, such as after hearing a news report of a fourteen-year-old killing a gang member, he'd grab my hand, look me in the eyes, and say, "I'm so proud of how much you read." Unfortunately, this pride never stopped him from grabbing the book out of my hands during events such as family dinners and church. Having to put my book away during family dinners really irked me because it was just posturing, I thought. There was no real reason for paying attention—no one was around except family, who didn't count as people. Sometimes my dad would let it slip. I'd make it through to dessert without anyone's mentioning the open book in my lap. However, if I missed someone asking to pass the milk or a question about my school day, my father would yank the book away, saying, "This is family time." I'd sneer at him and spend the rest of the meal limply eating, as if the forced abandonment of the fictional world had weakened me. Church was worse. I'd have to argue my case in hushed tones, pleading for my book back, knowing full well that there were no loopholes in Mass etiquette.

"There's a book right here," he'd say, and thump the Gospel. Indeed there was—the most boring book I'd ever read.

My grandmother saw a kindred spirit in me, giving me *Bleak House* by Charles Dickens from off her basement bookshelf when I was around thirteen. I think she glimpsed a John Grisham in my backpack and decided to set me right. It was the biggest book I'd owned up to that point. Sitting and reading it, I felt like I had finally arrived, an adult with a heavy hardcover. I used a bookmark—an accessory I usually deem unnecessary—despite my appreciation of folding pages. My grandfather, who passed on when I was fourteen, would always remind me to turn the light on if it were too dark while I was reading. He'd offer his oversized glasses to try on, the product of reading in too-low light, he claimed, in case I protested.

Grandma had my mother's sort of humbleness toward trivia and intellect, the "Oh, I don't know that answer" response, but, unlike with my mother, it was couched in an eagerness to share her speculations at the answer. While I was in college, surrounded by a bunch of snooty political theory and constitutional democracy majors who all lived with a singular devotion in life *to know everything*, it was my grandma we'd call when we got stuck on a clue in the crossword. She'd merrily tell me the answer, having finished it hours before. Sometimes she'd say, "Oh, you know this," and try to give me enough hints so I could maintain a bit of dignity by guessing correctly. Other times she'd concede that the question was a hard one and complete it for me.

We had a unique relationship, set by the fact that neither one of us was responsible for the other. That's the beauty of grandparents: all the fun with none of the blame. I told her my secret when I failed undergraduate precalculus for the second time, and she told me how her math teacher made her cry by accusing her of cheating after her father had shown her an easier way of doing long division. I told her when I tried smoking; she told me about how she used to go into the ladies' bathroom during lunch at work and practice smoking until she worked herself into a coughing fit, in order to look more professional like the male lawyers where she was a secretary. I went to her after I got sick from drinking for the first time and she made me laugh with a story about her drinking too much while eating chocolate at an awkward party. For every heartbreak I had, she had a story to match it. My grandmother was an amazing storyteller; I was in love with the way she could talk. I was obsessed with her stories, making her repeat them again and again. There was a comfort in the familiarity of her stories, how she made the same face at the same beat every time she'd tell the one about finding my mother as a toddler drinking dishwasher soap, how she phrased her description of the raw-egg-and-charred-toast-scrapings-filled drink she made to induce my mother's vomiting, and how she began to laugh before finishing when she described my mom's asking for more instead of vomiting as hoped.

My grandma was the light of my family's world. My grandfather was the backbone, a manly guide to football and how not to grill chicken. A provider and a punisher, the funniest and scariest part of every anecdote about my aunts' and uncles' childhood adventures. But my grandma was the softness, the solace—she was who we all went to when we had confessions; she knew all of our secrets.

Decades of smoking and an errant gene that, who knows, maybe I have as well gave my grandmother macular degeneration, a disease that slowly takes away the ability to clearly view objects right in front of you. The deteriorating eyesight blurs everything you directly look at, keeping peripheral vision fairly clear. As the disease progressed in her, her ability to read easily became less and less. She used bright lights and magnifying glasses to aid her, spending more time enlarging the words than actually enjoying them.

We helped her through surgery after surgery, fighting for her to keep the ability to read. Macular degeneration couldn't kill you, but losing out on reading was a fate worse than death to her. The last book my grandmother read, tellingly (in hindsight), was Christopher Buckley's *Losing Mum and Pup*. She was practically blind by this time; my uncle had given her a Kindle for Easter and we all sat around it after our brunch and set the options to the largest font size and highest brightness. That was the only book she bought on the device.

She lived for a year without the ability to read, a year spent in various hospitals, with cancer eating away at her bones. My grandma's cancer was slow and painful, a kind of death no one should endure. Particularly not people who light up their family's world.

After she died, I took the books from her "to be read" stack, a stack untouched since she left her home for the hospital. I dutifully read through them all and afterward placed each one on my bookshelf for books I want to display.

Why We Talk About It

YOU'RE AT A DINNER party and suddenly everyone delves into Twain. You sense the group is transformed by an instant connection, forged by the tradition of required reading in high school American literature classes. You can't help but weigh in with your own opinion on Tom Sawyer's funeral. In a more exasperating example of this phenomenon, two million reviews sprout up almost instantly every time a big-name contemporary author releases a book. Not many can stop themselves from saying something on the subject. But *why*?

Reading is a solitary activity. You can be surrounded by a thousand people, but processing the written words in your brain is something only you are going through. That's one of the reasons it feels so good to read on a park bench. Look up, people are walking by going about their days. Look down, the protagonist is confronting his family in a heated debate. Look up, the couple on the bench nearby are kissing. Look down, the hero is avenging his love. A good novel presents you with an engaging world that is a reality only for you.

A story is unbiased with respect to the reader. It presupposes nothing about the audience. Books don't require that you read them in a certain place, at a certain time, or with certain equipment. Just eyes. Literature connects by transporting people to the same consciousness; a stranger who's read the same book you've read, whose eyes passed over the same words, may be a part of a completely different environment, and even time, but for a while, at least, they shared a world with you. A community is built out of that isolated experience; an author has the power to build worlds and to populate them not only with characters but also with their readers.

While books are unbiased, the mark of a good one is that it makes some demands of you. Good books command study, presenting you with the puzzle of how and why their plot is laid the way it is laid—without examination the meaning is lost. In contemplating the answers to their questions, we collectively analyze in our book clubs, articles, and book reviews. A great author weaves a tale with a knot just loose enough to allow us to unravel it. A book read without deliberate consideration is a waste for both the author and the reader. Talking with others about our experience with the book is a way of celebrating the art. That we were pulled into the same existence and we emerged emotionally charged but each struck in a unique way is the beauty of reading.

So, we squeal the first time a new love tells us one of his favorite books is the same as ours. We bond with friends over plots that made us cry. We talk endlessly about how unlovable the author made a character. We fight over whether an ending was satisfying. We can't stop exclaiming when a book puts all the pieces together just right. We go on and on about the inexplicable phrasing of a character's emotion that changed our perspective on the feeling the next time we experienced it. We might not all agree on the nuances, we might emerge from the tale dissatisfied and jaded, but the joy of examining and debating our opinions is unrivaled.

The greatest argument for the oneness of humanity is the recognition that we are all emotional beings, subject to the fantasies of a story. We talk about this event we went through alone because it connects us together. You're never more human than when you realize a sentence has the power to push and pull the emotions of millions.

ACKNOWLEDGMENTS

THANKS TO MY EDITORS Michael Signorelli and Jason Sack for their patience.

Many, many thanks to my agent, Erin Malone, for her guidance.

Thank you to Patrick Moberg for his continuous, contagious creativity.

To my parents, thanks for all the books.

ABOUT THE AUTHOR

LAUREN LETO DROPPED OUT of law school after starting the popular website Texts from Last Night. She coauthored the book *Texts from Last Night: All the Texts No One Remembers Sending*. She grew up in the Detroit area but currently lives and works in Brooklyn.